Human-Environment Interactions

Mark R. Welford • Robert A. Yarbrough

Human-Environment Interactions

An Introduction

palgrave
macmillan

Mark R. Welford
Department of Geography
University of Northern Iowa
Cedar Falls, IA, USA

Robert A. Yarbrough
Department of Geology and Geography
Georgia Southern University
Statesboro, GA, USA

ISBN 978-3-030-56031-7 ISBN 978-3-030-56032-4 (eBook)
https://doi.org/10.1007/978-3-030-56032-4

structuresxx / shutterstock

This Palgrave Macmillan imprint is published by the registered company Springer Nature Switzerland AG
The registered company address is: Gewerbestrasse 11, 6330 Cham, Switzerland

Acknowledgments

We owe much gratitude and thanks to Rachael Ballard and Joanna O'Neill at Palgrave Macmillan in London for their patience and guidance throughout this process. We would also like to thank Brian Bossak who inspired Mark to think about writing a book on human-environment. We must also express our sincere thanks and appreciation for our families' support over the many years we worked on this project. We would not have been able to complete this book without the continued support of Theresa Welford, Mary Beth Yarbrough, Isabel Yarbrough, and Eleanor Yarbrough. Thank you, and we love you!

Contents

Introduction

Contents

M. R. Welford, R. A. Yarbrough, *Human-Environment Interactions*,
https://doi.org/10.1007/978-3-030-56032-4_1

Learning Goals

After reading this chapter, you will be able to:

- Define fundamental geographic concepts like globalization, spatial scale, and environmental determinism.
- Illustrate how fundamental concepts in geography aid in evaluating contemporary human-environment issues.
- Explain how the organizational structure of this book can contribute to your learning of contemporary human-environment processes, patterns, issues, and potential solutions.

Has our drive to build an advanced global civilization exhausted our planet's resources and polluted our planet to such a degree that we are in fact destroying the very civilization we are trying to build? Does our example explain why we have not located any evidence for life elsewhere? Are we an example of the Fermi paradox and the Great Filter where the Great Filter is an abiogenesis? In other words, is the rise of technological human-level intelligence ultimately self-destructive before we make contact with other alien civilizations? If so, then Gene Roddenberry, the creator of Star Trek, would be very depressed!

We hope not, and tragic major global crises such as the global economic crisis of 2007/2008, SARS, and COVID-19 suggest that downturns in the global economy yield positive environmental impacts! For instance, as of March 2020, nitrate oxide levels in China and within the Po Valley of Italy are down as much as 10–30%. As of March 27, congestion in and around Los Angeles is down and the traffic is moving 53% faster. Indeed, according to the Environmental Protection Agency's (EPA's) air quality index, by the end of March 2020, Los Angeles had recorded three straight weeks of "Good" air quality, which indicates little to no risks of air pollution. Contrast this with summer 2019, when EPA's air quality index in LA was in the "Unhealthy" range or worse every day for two straight months. Similarly, in the San Francisco Bay Area, air quality has improved markedly, with the number of vehicles crossing the Bay Bridge having dropped 40%, as has the number of vehicles driving into Seattle, Chicago, and Atlanta have seen similar trends, with massive downturns in numbers of vehicles on the roads.

Please note we do not advocate for emergent diseases and are horrified by the death toll of COVID-19.

Yet, this does illustrate that as people stay at home, consumption decreases, factory outputs decrease, transportation (including shipping, but also individual car drivers, trucks, etc.) decreases, and with this there is significant reduction in air pollution and greenhouse gas (GHG) emissions. We hope, from this evidence, that people (especially rich people) realize that reducing personal consumption (by a relatively small amount) can have an immediate impact on GHG emissions, and that we should, as a global society, explore zero-growth policies coupled with concerted efforts to reduce local, regional, national, and global socioeconomic inequalities.

As we noted in the last sentence, the onus is really on rich people to change their consumption patterns, particularly when it comes to flying. Oswald and coauthors

(2020) found that the richest 10% of people consume 20 times more energy than the poorest tenth in each region and across the globe. In fact, the richest 10% consume 187 times more fuel when traveling than the poorest tenth, wherever they live, and as they become richer, they use more energy in heating/cooling their homes, traveling, shopping, and eating. According to Harrabin (2020), in the United Kingdom (UK), 15% of flyers fly 70% of all flights, yet 57% of UK citizens do not fly abroad! Oswald and coauthors found that 20% of the UK population, 40% of the German population, and all Luxembourg citizens are among the top 5% of global fuel users. But, contrary to popular expectations, only 2% of the Chinese population and only 0.02% of the Indian population are among the top 5% of global fuel users.

In this chapter, we have several challenges—identify the scope of this project, from there define nature, define what are human-environment issues, define the Anthropocene, and begin to illustrate the connections between globalization, the defining attribute of our global society, and global environmental change, the defining by-product of our global society.

1.1 Anthropocene and Omnicide

Today, we live in the Anthropocene, an era dominated by humans, but also by the positive yet destructive environmental feedbacks that are poised to completely reset the relationships between nature and society! But what is the Anthropocene?

» With world population projected to reach at least 9 billion by 2050, human interactions with landscapes are increasing at unprecedented rates. Indeed, the environmental impacts of human population growth and accompanying resource consumption have intensified to the extent that the term '**anthropocene**' has emerged to signify a new geologic era dominated by human activity.

» The Future of Human-Landscape Interactions (AAG Newsletter, Dec 2010), Ann Chin and Carol Harden

> Temporally constraining the Anthropocene has proved difficult. According to Waters et al. (2018), the Anthropocene Working Group representing the American Geosciences Institute and earth scientists worldwide is suggesting the "golden spike" in radioactive nuclear fallout and changes in carbon chemistry through fossil fuel burning between 1952 and 1955 should stratigraphically define the new Anthropocene epoch. Carrington (2016) suggests that other proposed Anthropocene stratigraphic signals might include the presence of plastic pollution, coal soot from power stations, concrete and concrete dust, and domestic chicken bones in rubbish dumps and kitchen hearths.

Chin and Harden continue by posing a series of questions related to the human and physical dimensions of landscape change, including:

1

» What are the unintended and broader social and political consequences of landscape change, especially when a small change in landscape processes can produce large social change, and vice versa? What common currency can express the value of environmental goods and services, and relate environmental systems to economic and political systems?

While agreeing with their premise, we take issue with the authors' privileging of population growth and their omission of the immense diversity of resource consumption and technological access across the globe. However, we would be derelict if we did not mention that between 1990 and 2010, globalization, capitalism, and market socialism (China's model of growth) helped more than a billion people out of poverty. But these three systems might have reached their zenith. Increasing use of technology in the Global North in manufacturing, product delivery, and services will likely cause 47% job losses among the middle and working classes in the next 25 years according to Frey and Osborne (2017). This at a time when middle- and working-class wages have stagnated; for instance, 15% of the US population today exists below the poverty-line, while all productivity gains have accrued to the top 1% of earners.

However, the utilization of space by humans or the lack of space for other species is certainly an issue. Sir Richard Attenborough states, "space is not as sexy as plastic, it's a harder thing to get your head around, it's a much bigger issue." For instance, blue whales are struggling to find partners because the noise of the oceans is drowning out their songs, so their ability to communicate and find each other over vast distances is diminishing. By 2050, 10% of all animal and plant species or a million species are expected to go extinct according to Chris Thomas and coauthors (2014). Among the first to go are those animals larger than 100 kg; these include elephants, rhinos, giraffes, lions, tigers, many whale species, blue-fin tuna, orangutans, and most shark species, among many others. All these species require significant landscape or space to survive. Today's current extinction rate is at least 1000 times faster than the background extinction rate since the Chicxulub event 66.043 ± 0.011 million years ago.

Yet space is not the only pressing issue. Business-as-normal is consuming all earth's natural resources (i.e., minerals, timber, soil, freshwater and groundwater, clean air) and generating enormous pollution, of which one form, GHGs, is conspiring to unnaturally warm our earth to levels not seen in millions of years. This at a time when urbanization is accelerating! And yet according to Felix Creutzig and coauthors (2015), urban areas consume more energy, spread more diseases, consume more water, and generate more trash and more GHGs than rural environments.

Both Frank Fenner and Jared Diamond liken our spiraling descent into chaos triggered by global warming, business-as-usual resource consumption, and loss of natural landscape space as reminiscent of Easter Island. Both are worried that humans "lost in space" will perish! Both are worried that competition for the last remaining natural resources, for example, water and soil, will trigger wars over food and water!

Furthermore, we would augment these passages by noting the severity of humans' impact on the environment, where some instances of environmental degradation have reached their thresholds, such that the extent and severity of the impacts may prove to be irreversible. In addition, physical and human-driven processes work in tandem to produce the constantly changing landscapes that Chin and Harden identify, and the scope and extent of such drivers remain difficult to predict precisely because of such complex interactions. Moreover, those processes operating at a global scale result in geographically uneven consequences for both societies and physical environments, such that places and peoples neither contribute to nor are affected by environmental impacts equally.

Although the term Anthropocene captures the essence of our era, it tends to camouflage a sinister reality—we are all of us, at different scales and different rates but especially Global Northern citizens, culpable of "ecocide" or rather "omnicide"! We prefer the term omnicide to ecocide, the killing of ecosystems, as ecocide is spatially more restricted in interpretation.

> Omnicide—the killing of everything is global in its reach. We have actively and passively conspired to create conditions in which our modern highly interconnected, highly wasteful, highly resource-exploitive global civilization is initiating the sixth mass die-off or sixth mass extinction. As Danielle Celermajer suggests, we are not just killing animals, trees, fungi, biomes, forests, and rivers, we are also killing humans through pollution and natural resource exhaustion.

According to Danielle Celermajer, and we concur, we need to identify the political representatives, the media representatives, the financial institutions, the businesses, the governments, and individuals who are, at whatever geographic scale, facilitating and are culpable for omnicide of the earth. We as citizens and as scientists must not hide or be quiet anymore; we must remember Elie Wiesel's quote about the Holocaust and apply it to our modern context and change it subtly:

» We must always take sides. Neutrality helps the polluter, the natural resource exploiter never the natural environment. Silence encourages the environmental abuser.

This is why environmental abusers, governments and businesses alike, target the likes of Sweden's teenage climate activist Greta Thunberg and others.

Thus, a more explicit integration of human-driven and physical processes coupled with careful attention to issues of geographical scale, environmental thresholds, and geographical unevenness potentially provides a comprehensive and sophisticated depiction of human-environment interactions in the twenty-first century. Such an integrative approach that draws on fundamental geographic concepts to represent and analyze the complexities involved in the production and impacts of human-landscape processes that mosaic the earth's surface provides a framework for this book.

1.2 Geographic and Environmental Concepts

At first glance, "nature" may seem a fairly obvious, self-evident concept that lends itself to a simple definition.

> Nature is frequently defined or understood differently across societies or cultures, leading to a conception of nature as socially constructed. Such a perspective highlights the ways in which a given society's views of nature impact issues of resource management, development, environmental policy-making, and a whole host of other components that shape the human-environment mosaics that pattern the globe.

Moreover, fundamental concepts from geography such as geographical scale, boundaries, and environmental determinism are essential to building a knowledge base necessary to critically analyze human-environment interactions.

Within geography, intellectual spaces and subject domains have been articulated around nature-society and human geography, and human-environment and physical geography, with some overlap with GIScience.

> The principle themes associated with human-environment studies are the dynamics of coupled human-environment interactions, land-use and land-cover change, and the resilience and vulnerability of social-ecological systems. Whereas the principal themes of nature-society studies (NS) involve environmental governance, environmental history, and environmental management and policy construction.

There is a growing consensus among social scientists that all knowledge is produced within distinctive social and cultural contexts that invariably influence and condition the process, and by extension the application, of new knowledge in policy arenas. The Aboriginal Indigenous Engagement Model conceptualized and implemented by a consortium of the CRC Plant Biosecurity (PBCRC), now The Australian Plant Science Foundation (APBSF), The New Zealand Institute for Plant and Food Research, and Charles Darwin University is just such an attempt to involve social and culture contexts in policy engagement and environmental management. This model draws parallels between the traditional processing of cycad seeds and effective and inclusive community engagement.

Furthermore, in the past, natural scientists and/or environmentalists used their privileged positions to propose environmental policies with little regard to the cultural, social, political, and economic characteristics of human societies. Recently, this has begun to change. For example, since 1998, the American Association for the Advancement of Science has supported interdisciplinary research as the means to solve the complex problems that earth faces in the twenty-first century. Natural scientists have been, until recently, less inclined than social scientists to engage with this approach, focusing instead on collaborative environmental assessment projects undertaken with governmental and private agencies. The recommendations and outcomes of such projects often face fierce criticism because of divergent viewpoints over the appropriate management of natural resources and landscapes. Perhaps of most concern, the public often fails to appreciate the scientific culture within which such projects are conducted. It is our contention that geographers, landscape and political ecologists, and political economists, among many others, can begin to overcome this unfortunate disconnect between the public and natural science arenas by applying concepts from multiple social science disciplines to engage the public in a more diverse and reflexive form of scientific practice and environmental management. It is within this academic and social context that we address today's complex mosaic of human-environment interactions.

Fundamentally, the interactions among humans and the environment occur across space and time. Thus, the resulting rich and spatially varied mosaic of human-landscape products we see today exists as the sum of modern and historic political, social, and cultural processes and physical landscape responses. Things happen, they happen across the earth's surface, and humans both affect and are affected by physical landscapes. Yet different cultural and social groups, political and economic entities view, react to, and thus impact the human-environment processes differently. Therefore, although historical legacies of past human and environmental interactions remain today, it is the spatial representation of these historic interactions and the extant contested interplay between humans and nature that lie at the heart of this relationship. Thus, our interdisciplinary approach to writing this book involves positioning political ecology within a spatial dimension.

1.3 Nature as Socially Constructed

» "It is a paradox of our modern time that" the "immense success of science is being questioned... Science is faulted for its privileged status"; moreover, "scientists are subject to the same pragmatic principle that governs other members of society: It does not matter whether an idea or concern of people is real or not; if people believe it is real, then it will be real in its consequences. The present state of 'science literacy' among the potential believers of our society does not give comfort that their reasoned understanding of science will thwart any adverse consequence of current antiscientific trends." Baker, V.R. "Let Earth Speak," pages 359–367 in *The Earth Around Us: Maintaining a Livable Planet*, edited by Jill S. Schneiderman. W. H. Freeman and Co. (2000).

This warning, equally concerning to physicists, biologists, atmospheric scientists, geologists, and physical geographers, is very apt today. Yet this warning should be extended to include a warning about deconstruction theory and its use to devalue or reject scientific methodology, scientific results, and recommendations originating from these results.

1

> *Deconstruction theory* suggests that any given text has irreconcilably contradictory meanings, rather than being a unified, logical whole.

Deconstruction criticism of science is alarming, but it is particularly worrying that most climatologists, physical geographers, physicists, chemists, and geologists are neither aware of this issue nor have seemingly much to offer to rebuff deconstruction criticism of their work. It is all the more worrying because, in recent years, Nellis argues that geographers (both physical and human) have been called to "play a more active role in information management at all levels and related geographic policy," and as Richardson notes, geographers have been encouraged to "emphasize unity, inclusion and the common ground within our discipline." Richardson goes further and states that "We must learn to work effectively within groups with divergent viewpoints, and to focus on consensus building and inclusion of disparate perspectives."

Within biology, Soule and Lease have tried to alert biologists to the severity of the problem presented by deconstruction theory and its application to the biological sciences. All physical and social scientists cannot simply ignore these threats; we must understand the nature of these threats, develop arguments to counter them, and be ready to counter deconstructionist arguments to college administrators and the public. We cannot continue to keep our heads firmly embedded in the sand, even if the sand does have some rather interesting and unusual cross-beds in it!

1.4 Semantic Challenges to Nature and Society

Beginning in the late twentieth century, postmodernists levied criticism on the entire range of practices and epistemologies (i.e., means of knowledge production) in academia. Some fields of study, however, were quicker to respond to these criticisms than others. Scholars in cultural studies, literary criticism, sociology, and ethnography attempted early on to integrate postmodern ideas into their fields, which led to new, stimulating arguments regarding the production of knowledge and the importance of alternative rationalities. Natural scientists, by contrast, have been much less inclined to engage in the debate, although a number of biologists recently generated an articulate first response to postmodern assessments of scientific inquiry. Physical geographers have a tradition of privileging fieldwork and codified methods of analysis over reflection and philosophical exploration. With a few notable exceptions, the subfield has simply chosen not to participate in the discussion and focus instead on collaborative environmental assessment projects undertaken with governmental and private agencies. The recommendations and outcomes of such projects often face fierce criticism because of divergent viewpoints over the appropriate management of natural resources and landscapes. In addition, the public often fails to appreciate the scientific culture within which the projects are conducted.

Postmodernism is not a singular critique or philosophical position. Indeed, some have argued that it is inappropriate to use the broad label "postmodernism" at all because its practitioners resist all forms of classification. Clearly, though, several thematic patterns can be said to characterize most postmodern literature.

> At root, postmodernism is a critique of language. Modernity assumed a phenomenological separation of subject from object, mind from matter.

The natural world, moreover, possessed an intrinsic order that could be discovered through the deductive process of observation, abstraction, empirical testing, and theory construction. In all of this, it was assumed that the system of abstraction—language itself—was stable, fixed, and consequently reliable. Indeed, a highly refined system of abstraction, mathematics, was always seen as the essential tool in deciphering the intrinsic order of the natural world.

> Postmodernism, by contrast, denies the possibility of a singular truth. Its practitioners underscore the unstable and accidental nature of the relationship between signifier and signified. The meaning of a word does not depend upon the external object it signifies, but rather exists only in reference to other words and signs in a chaotic and ultimately incomprehensible relationship. Language, then, cannot provide direct or unproblematic access to reality, but rather *shapes* reality within the confines of its own conventions and practices; all independent of a postulated "real" world "out there."

Consequently, the entire Cartesian dichotomy between a rational subject and an external, but ultimately knowable, world is pure illusion. The individual is the unstable product of a hyper-relative practice of signifying an infinite regress; most postmodernists question the coherent individual and prefer to speak of multiple subjectivities or polymorphic subjects. The convention of "truth" is society's way of stopping the infinite regress of language to make sense of the world, if only partially and temporarily. Postmodern critics readily observe, however, that modern society universalizes such half-truths into absolute statements. The production of truth, then, requires the constant application of social power to silence alternative views and to make linguistic apparitions appear as unquestioned common sense, a taken-for-granted dimension of everyday experience. For this reason, postmodernism conceives of truth as a fiction backed by power, a form of terrorism designed to maintain the position of those whose interests it serves.

Most of these assertions anger physical scientists, whose work is premised on the practice of generating valid and reliable knowledge about the natural world. The tendency is to dismiss the entire critique as specious and unfounded. How can postmodernists make *assertions* about the illusory nature of truth when they them-

1

selves contend that language prohibits the very act of making such strong declarations? But, the postmodernist case is much more nuanced than usually perceived. In fact, it is possible to recognize two broad, although not exclusive, camps: *skeptical postmodernists* and *affirmative postmodernists*.

> Skeptical postmodernists are known for the outrageous arguments outlined earlier regarding the hyper-relativity of knowledge and the terrorism of truth. Many skeptics would renounce the validity (and authorship!) of their own work, and in fact routinely make contradictory arguments, sometimes within a single text or interview.

Some of the most extreme skeptics have written themselves into obscurity and disdain, most notably the founder of the technique of deconstruction, Jacques Derrida.

The affirmative position, by contrast, offers much more promise for positively modifying the processes whereby knowledge is produced, in order to better incorporate diversity, innovation, and alternative viewpoints. The affirmative view maintains a basic form of the earlier argument that pure, universalized truth is impossible, and that truth claims are made by marshaling and appealing to the structures and relations of social power. But affirmative postmodernists refuse to accept the idea that *all* knowledge is *completely* fallacious, and that we can never know *anything*.

> Affirmative postmodernists generally contend that all knowledge is produced within distinctive social contexts that invariably influence and condition the process.

So, although it is inappropriate to seek absolute, universal truth, it is possible to create partial, diverse, and conditionally valid truths that, taken together and played off one another, provide a rather generous and ultimately collective understanding of our social and physical environs.

1.5 Spatial Scales and Boundaries

The adjective "spatial" comes from the root word "space," and spatial scale is a fundamental concept in the discipline of geography with myriad definitions and conceptualizations. Indeed, academic geographers have debated the conceptualization, origin, and utility of spatial scale for decades, and many of those conversations continue today. Notwithstanding these ongoing discussions, spatial scale remains a useful concept in examining human-environment interactions, environmental issues, and environmental policies.

Spatial scales are a means of organizing geographic space, frequently by dividing it into levels or portions of varying size. For example, the global scale refers to the entire earth, including all physical processes (e.g., surficial and atmospheric) and human-driven processes (e.g., demographic, cultural, and economic) contained therein.

For our purposes, the global scale would be the largest, most expansive spatial scale, within which we might identify smaller portions of geographic space (i.e., smaller spatial scales). World regions, such as East Asia, sub-Saharan Africa, and South America, for instance, exemplify a different spatial scale than the global since they refer to only a portion of the entire earth. As we continue to "zoom in" from the global scale (like you might do on your mobile phone), we alter the spatial scale at which we are viewing the earth. Sticking with the zooming in metaphor, these smaller and smaller portions of the earth enable us to see more detail, all the way down to the household scale or (theoretically) the scale of the body (although your phone app obviously is not powerful enough to zoom in that far). Thus, spatial scale is a way of conceptualizing and organizing geographic space as we zoom in or zoom out. Common examples of spatial scales would include the following (although this is by no means an exhaustive list): global, world regional, national, intra-national/regional (e.g., the northern India or the US South), urban (i.e., the city), neighborhood, household, and the body. Note that each one of these examples of spatial scales is a smaller area than the previous one, as we are continuing to "zoom in" on the earth.

Geographer Andrew Herod has written extensively on spatial scale, and his work illustrates the common understanding of the origin of spatial scale. Herod's research, and that of many other geographers, demonstrates that spatial scales are socially constructed and dynamic rather than natural and fixed. What does it mean to say that spatial scales are socially constructed? A historical example might help illustrate. Today, we refer to the national scale to represent processes originating from and/or affecting a country. For instance, when we compare annual greenhouse gas emissions of the United States to those of China, we are examining data at the national scale (note that although China emits more greenhouse gases each year than does the USA, this is a function of population. On a per capita [i.e., per person] basis, the USA is a much bigger emitter of GHGs than China). Yet, for most of human history, there was no such political entity as a "country," and thus no such thing as the national scale (think, the Roman Empire, the Mughal Empire that dominated current-day South Asia throughout much of the sixteenth to nineteenth centuries, or even the imaginary kingdoms in *Game of Thrones*). Countries have become such a common means of organizing space on the basis of sovereign political control, that they may seem natural and concrete. However, the concept of the country (or state in political geography terms) has only existed for approximately 400 years, a long time to be sure, but only a small piece of human history. We can say the same thing for the urban scale or the spatial scale of the neighborhood. Not all cities or neighborhoods are the same size and they do not stay the same over time. This illustrates how spatial scales are socially constructed and impacted by geographic and historical contexts; that is, what counts as the urban, neighborhood, or national scale (or whether these even exist!) depends on when and where we are. Spatial scales, therefore, are not an arbitrary, artificial framework for dividing up geographic space that is simply overlaid on the earth's surface, like a grid of smaller and smaller squares that might be placed on a globe. Instead, spatial scales are defined and delineated according to human or physical processes that exist in the world and act upon space (e.g., neighborhoods, cities, regions, or countries).

In addition to recognizing that spatial scales are dynamic and fluid, some familiarity with a few metaphors is helpful in understanding how spatial scale is utilized throughout this book. The first metaphor is that of a ladder, wherein larger scales are the top rungs of the ladder (global, world regional, national, etc.) and spatial scales get smaller as we move to a ladder's lower rungs. This metaphor is helpful in some ways and misleading in others. One critique is that it fails to highlight the interconnections between scales, while also privileging the larger scales over the smaller (since they are higher up on the ladder). Think for a moment about your neighborhood; notwithstanding its uniqueness, is it not still part of the urban, regional, and national scales? Of course it is; we cannot separate it from the larger spatial scales within which it is apart. One spatial scale is not distinct or separate from other scales, but embedded within them. Hence, a concentric circles metaphor (one circle surrounded by a series of larger and larger circles) may be more helpful in envisioning the concept of spatial scales. One final metaphor that geographers turn to regarding scales is that of matryoshka dolls (more commonly known as "Russian nested dolls"). These wooden dolls are a series of identical dolls of various sizes, such that one fits inside another and those fit inside a larger doll until all of the dolls fit inside the largest one. This nested doll metaphor is helpful in reminding us that all spatial scales from the local to the global are intimately connected to each other. It is also worth noting that the smallest doll (i.e., the smallest scale) is not necessarily the least significant.

Related to socially constructed spatial scales, the term endemism illustrates the confusion that human-delineated spaces versus more organically derived space can present.

> An endemic is broadly defined as range-restricted species.

Until recently, endemics were identified at the national scale—for instance, Ecuador has 6 bird species restricted to mainland Ecuador and 30 restricted to the Galapagos Islands. However, Birdlife International introduced a more inclusive and ecologically appropriate term, Endemic Bird Areas (EBAs). EBAs contain habitats where multiple restricted-range bird species (those species restricted to less than 50,000 km^2) uniquely occur. The Chocó EBA that includes portions of Ecuador, Colombia, and Panama has 62 endemic bird species, and to illustrate the EBAs greater worth, 210 endemic amphibians, 63 endemic reptiles, and maybe as many as 1600 endemic plant species.

Spatial scales and human boundaries are also not necessarily restrictive. Cultural and socioeconomic processes frequently jump scales, such that local environmental issues can become regional or national or international. The 1997 proposal to create another oil pipeline across the Andes, from the Amazon to the Pacific in Ecuador, mobilized local activists who sought help from regional, national, and international organizations to first stop the pipeline, and then reroute it away from the Chocó EBA and the newly identified Mindo Important Bird Area (IBA). The Mindo IBA was created by Birdlife International as a reaction to the proposed pipeline route. The *San Francisco Chronicle* even carried an article titled *Pipeline through paradise. Oil route across* Ecuador's *cloud forest threatens birds -- and ecotourism*. Ultimately, the pipeline was finished in 2002, but Mindo became even more well known. As one long-time environmental activist explained, "A bad thing became a good thing." He continued, "Mindo became a model for Ecuadorians to conserve nature, a model about birding, a model about how local people can work on these issues---you can ask everybody in town and everybody can tell you something about conservation."

Thus, spatial boundaries can be at the forefront of environmental conflicts as another recent historical example illustrates. In 1975, Iceland established a 200-mile exclusion zone to try to conserve their cod fishery. The UK responded that all ocean resources 3 miles beyond any coastline are a common resource to be shared by all. The UK sent Her Majesties' Navy to protect the rights of UK fishing vessels within the 200-mile zone. Iceland responded by arming coast guard vessels with fishing trawl cutters and cutting trawler nets and firing across and in front of trawlers to try and turn them back. The Cod War ultimately led the UN to embrace a 200-mile exclusive economic zone for all countries possessing an ocean coastline.

In conclusion, we employ the concept of spatial scale throughout this book to assist in achieving a number of goals. Among these are using spatial scale to examine ecological processes and environmental impacts across space, to illustrate the "mismatch" between spatial scales socially constructed by human activity and spatial scales constructed via ecological or biological processes (e.g., the Endemic Bird Area), and to highlight examples of how "jumping scale" or "scaling up" have affected environmental movements and/or policies.

1.6 Environmental Determinism

This deeply racist philosophy argued that the physical environment predestined societies toward particular socioeconomic and cultural trajectories. In short, physical characteristics, especially climate, *determined* culture and society.

This deterministic approach is a far cry from the suggestion that the physical environment in some way shapes culture (e.g., climate and soil types impact agricultural practices).

The idea that nature affects culture is an illustration of possibilism; that is, given a certain set of environmental factors, a multitude of cultural outcomes is possible. Thus, possibilism should not be confused with determinism.

In a nutshell, environmental determinism concluded that mid-latitude environments with their highly variable weather created a rich learning environment, while polar and tropical environments with their highly predictable and unvarying climates did the opposite. This philosophy was a product of Judeo-Christian and colonialist mentalities, and it has no basis in fact.

1.7 Environmentalism

In a modern sense, the battle between Gifford Pinchot (first chief of the US Forest Service) and *conservation* and John Muir and *preservation* in the USA led Aldo Leopold to consider a more inclusive land ethic. This is widely believed to have heralded the beginning of environmentalism. Leopold thought that ethics direct individuals to cooperate with each other for the mutual benefit of all. He argued that this "community" should be enlarged to include nonhuman elements such as soils, water, plants, and animals, "or collectively: the land." Although the release of the 1949 Sand County Almanac launched the land ethic, the modern contemporary American environmental movement did not emerge until Rachel Carson published *Silent Spring* in 1962 that identified to a broad audience the link between DDT and harmful trophic cascades. The *Tragedy of the Commons* published by Garrett Hardin in 1968 identified where individuals using a common resource for their own personal gain degrade the common resource, leading to a decrease in yield for both the group and the individual. Among the many tragedy of the commons examples are tuna fishing, point-source polluting of air and water, and utilization of common pastureland. In the 1970s, *Animal Liberation* by Singer challenged human use of animals for pharmaceutical testing and human consumption, while Næss' *Deep Ecology* established the contemporary ecological philosophy that all living things have inherent worth, while emphasizing the interdependence of organisms within ecosystems. Today, ecosystem services and ecofeminism are the buzzwords of environmental movements.

Ecosystem services are those cultural-socioeconomic benefits that people obtain from ecosystems. Without support, provisioning, regulating, and cultural services from ecosystems, humans and our civilization would cease to exist.

Ecofeminists argue Western culture oppresses and subordinates both women and nature, suggesting men rank higher than women and culture is more important than nature.

Approach

The book is divided into three broad sections: Section I, the Issues and Challenges of Climate Change, threshold exceedance and the sixth mass extinction; Section II, the Settings that contribute to Section I—population, agriculture, and resource depletion; and Section III, our take on Practical Solutions to our Human-Environment Crises. After providing students with some fundamental tools in geography and environmental science, the text moves to an examination of several issues and challenges that have arisen as a result of human-environment relations and concludes with a discussion of multiple approaches to addressing the challenges highlighted in Sections I and II. Our approach to teaching about human-environment relations draws on the synergistic nature of this field in the organizational framework of the text.

As a specific organizational theme, we interweave discussions of human-driven processes and physical processes on the earth's surface within each chapter (as opposed to having separate human and physical sections). As such, the organizational framework of each topical chapter will reflect the ways in which human- and physical-driven processes affect and are affected by one another to create the world in which we live. The final section seeks to synthesize and apply information presented in previous chapters to offer analyses of contemporary human-environment issues (e.g., water resources, climate change, and food security). Our hope is that these analyses provide instructors with a means of teaching students how concepts in Section I of the text are applied to issues and challenges in Section II of the text, oftentimes through a synthesis of multiple environmental issues. Thus, Section III will also offer instructors a means of evaluating the degree to which students can analyze and evaluate contemporary environmental challenges through the application of fundamental human-environment and geographic concepts.

1

References

Carrington, D. (2016). The Anthropocene epoch: Scientists declare dawn of human-influenced age. *The Guardian*, 29. https://www.theguardian.com/environment/2016/aug/29/declare-anthropocene-epoch-experts-urge-geological-congress-human-impact-earth. Accessed 23 Jan 2019.

Creutzig, F., Baiocchi, G., Bierkandt, R., Pichler, P. P., & Seto, K. C. (2015). Global typology of urban energy use and potentials for an urbanization mitigation wedge. *Proceedings of the National Academy of Sciences, 112*(20), 6283–6288.

Frey, C. B., & Osborne, M. A. (2017). The future of employment: How susceptible are jobs to computerisation? *Technological Forecasting and Social Change, 114*, 254–280.

Harrabin, R. (2020). *Climate change: The rich are to blame, international study finds*. BBC News. https://www.bbc.com/news/business-51906530

Oswald, Y., Owen, A., & Steinberger, J. K. (2020). Large inequality in international and intranational energy footprints between income groups and across consumption categories. *Nature Energy, 5*, 231–239. https://doi.org/10.1038/s41560-020-0579-8.

Thomas, C. D., Cameron, A., Green, R. E., Bakkenes, M., Beaumont, L. J., Collingham, Y. C., Erasmus, B. F., De Siqueira, M. F., Grainger, A., Hannah, L., & Hughes, L. (2004). Extinction risk from climate change. *Nature, 427*(6970), 145.

Waters, C. N., Zalasiewicz, J., Summerhayes, C., Fairchild, I. J., Rose, N. L., Loader, N. J., Shotyk, W., Cearreta, A., Head, M. J., Syvitski, J. P., & Williams, M. (2018). Global boundary stratotype section and point (GSSP) for the anthropocene series: Where and how to look for potential candidates. *Earth-Science Reviews, 178*, 379–429.

Further Reading

Celermajer, D. *Omnicide: Who is responsible for the gravest of all crimes?* ABC Religion & Ethics. https://www.abc.net.au/religion/danielle-celermajer-omnicide-gravest-of-all-crimes/11838534

Cotgrove, S., & Duff, A. (1980). Environmentalism, middle-class radicalism and politics. *The Sociological Review, 28*(2), 333–351.

Domonoske, C. (2018). *In March, Portugal made more than enough renewable energy to power the whole country*. NPR. https://www.npr.org/sections/thetwo-way/2018/04/05/599886059/in-march-portugal-made-more-than-enough-renewable-energy-to-power-the-whole-coun. Accessed 17 Sept 2018.

Harrington, B. (2019). *'Aristocrats are anarchists': Why the wealthy back Trump and Brexit*. Guardian. https://www.theguardian.com/us-news/2019/feb/07/why-the-wealthy-back-trump-and-brexit?CMP=Share_iOSApp_Other&fbclid=IwAR2tu_4R43IttfYHDTI98vnKjYSqMSEDsALo6nrWqIV1GeNpFlIznVfF9Rw. Accessed 7 Feb 2019.

Hirsch, F. (2005). *Social limits to growth*. London: Routledge.

Turner, G. M. (2008). A comparison of the limits to growth with 30 years of reality. *Global Environmental Change, 18*(3), 397–411.

2 Climate

Contents

M. R. Welford, R. A. Yarbrough, *Human-Environment Interactions*,
https://doi.org/10.1007/978-3-030-56032-4_2

2

Learning Goals

After reading this chapter, you will be able to:

- Expound on the cause of global warming.
- Articulate the ways in which we have exceeded critical earth-atmospheric thresholds and how global temperatures point to acceleration in global warming.
- Explain why there is no comprehensive global-wide policy on climate change.
- Illustrate how politicians and business leaders in the Global North simply are not prepared to reduce $C0_2$ production.
- Appraise claims that deny anthropogenic global warming and climate change.

2.1 Climate Change Is a Game-Changer

Climate change is not an esoteric "them-not-us" issue: it is a fundamental game-changer for the modern standard of living that the Western world has become accustomed to over the last century of development. We would be ill-advised to forget that the Mayan and Anasazi civilizations and Norse Greenland islanders were felled by climate change, coupled with carrying capacity exceedance, although diseases could have also played a knockout role.

According to National Oceanic & Atmosphere Administration (NOAA), January 2020 was the hottest January on record, 2.05 °F above the twentieth-century average January temperatures. On February 9th, the Antarctic experienced its first temperature that exceeded 20 °C or 68 °F. ***This is not normal!*** In Europe, according to the European Union's Copernicus Climate Change Service (C3S), the months of December (2019), and January and February in 2020, were the hottest on record, 1.4 °C above the winter record set in 2015–2016. Incredibly, Helsinki registered 6 °C higher than normal for January and February. In February 2020, repeated intrusion of warm, humid tropical air brought very wet weather to the United Kingdom (UK). The UK's average precipitation for the month was 237% above average or 209.1 mm of rainfall.

The IMBIE team (a team of 89 scientists who work on the Ice Sheet Mass Balance Intercomparison Exercise) calculated that in the 1990s, Antarctica and the Greenland ice sheets lost 81 billion tons per year, but in the 2010s, this rate of loss had risen to 475 billion tons per year. Team leader Andrew Shepherd and the IMBIE team suggest that if this rate of melting continues through the end of the century, sea-level will rise 17 centimeters, placing 400 million people at risk of increased coastal flooding. Those cities most at risk include Venice, Italy; Charleston, South Carolina; Miami, Florida; Jakarta, Indonesia; Amsterdam and Rotterdam, the Netherlands; and London, among many others! Some island countries have very difficult decisions to make in the near future: (1) they either have to build higher, very expensive coastal levees, (2) dredge up reef material from their lagoons and cover their entire islands to raise it, or (3) abandon their islands for higher ground elsewhere if this is at all possible. These include the Grand Caymans, Tuvalu, and Majuro in the Marshall Islands, and the Federated States of Micronesia, among others. According to Velicogna and coauthors, in just two summer months in 2019, the Greenland ice sheet lost 600 billion tons of ice, resulting in a 2.2 mm rise in global sea level. This suggests that if global warming continues to accelerate, the IMBIE projections will seriously underestimate future sea-level rise! (◘ Fig. 2.1)

If we step back in time—the Western world was recently rocked by climate change and a global pandemic between 1346 and 1815 when the Medieval Black Death (MBD) ravaged Europe and Europe was caught in the grips of the Little Ice Age. Together they killed millions! The primary wave of the Medieval Black Death in 1346 was preceded by an enormous famine in 1315, suggesting that agricultural technologies and population-carrying capacities were at a breaking point. At this point in history, the prior Medieval Warm Period (950–1250) was ending due to the failure of the deep Atlantic Conveyor and North Atlantic drift, as freshwater flowing off the melting Greenland ice cap halted the sinking of hyper-saline waters off the coast of Greenland. According to Van Hoof and coauthors, the Black Death pandemic appears to have fueled this cooling as millions of acres of farmland underwent secondary succession back to forest trapping and sequestering millions of tons of CO_2. The primary wave killed 35–55% of the European population, and by the time the medieval Black Death disappeared, as many as 75 million had died.

The millions in Europe who survived this period from 1347 to 1815 witnessed and participated in some of the greatest changes in human history: the Renaissance, the Enlightenment, the British and French Revolutions, and both the Agricultural and Industrial Revolutions that initiated the transition from animal power to steam

◘ **Fig. 2.1** A nearly ice-free Northwest Passage, August 9, 2013. NASA Earth Observatory

2

power, and the move from wood to coal energy resources, and the rediscovery of concrete. Europe went from the Dark Ages to the Age of Colonialism, the precursor to our system of multinational corporations and globalization. But it is important to note that the immediate period after the primary wave of the Black Death was racked by widespread violence, lawlessness, widespread distrust of church hierarchies, and political upheaval. But even here, feudalism died, while towns and cities and their gilds, the precursors to modern corporations, grew, and land was swapped for money as a sign of wealth. Many rich families disappeared, and while the poor suffered the highest mortalities, some survivors from the rural and urban underclass rose to wealth and prominence. In fact, Tom Nicholas suggests the British class system, rather than being immanent, is highly fluid, suggesting the rise and fall of families across generations is the norm—wealth, power, and privilege, in fact, rarely seem to survive more than three generations.

So, while we sit comfortably reading this text, let us remember that environmental perturbations, be they either disease, climate change, or resource exhaustion, have precipitated huge human die-offs and radically changed civilizations. But economic or environmental perturbations can also provide *opportunities* to groups who were previously discriminated against, or who were isolated from the benefits of wealth, trade, or education.

Case Study: EU

Although the picture admittedly appears bleak at first glance, we posit that such challenges are not unprecedented and that deliberate techniques have worked to prepare for and respond to such civilizational challenges. A good example is Europe's rather belated response to the world conflicts that erupted twice on that continent. Following World War I, Europe and the rest of the world tried to develop cooperation across divided Europe to forestall any future conflict. However, Shirer suggests the League of Nations had no real power and the allies were bent on extorting Germany for war reparations, ultimately bankrupting Germany and leading to the rise of Nazism and Adolf Hitler. After their miserable post-WWI failure, Germany, Belgium, Italy, and France were in no mood to replicate these events following WWII, and set Europe on the course to the European Economic Community (EEC) and the European Union. In particular, Jean Monnet, architect of the European Coal and Steel Community (ECSC) and EEC, understood that fundamental and transparent social, economic, and political cooperation across nations stood at the foundation of any attempt to survive or avoid any political, economic, or environmental disaster. Monnet persuaded six governments—Italy, West Germany, France, Belgium, Luxembourg, and the Netherlands—that they must adopt a radical new approach to European politics: cooperation, not confrontation. So far, Monnet's creation has, at least within Europe, stopped another continent-wide conflagration. In a sense, the EU saved western European civilization. Today, according to Rifkin, the EU is the largest economy on earth. It also has the strictest environmental regulations and sponsors the broadest conservation measures across the planet. It legislates to maintain public transport, funds the development of high-speed rail connections, and makes many other attempts to reduce community member CO_2 emissions. Although there is a siren call to peoples across the world to "think global, act local" to address issues of rampant, unsustainable, unethical globalization and resource depletion, at least the EU might offer EU member-states and citizens an organization possibly capable of coordinated planning and preparation for Kunstler's "Long Emergency" at the supra-national level.

Nevertheless, the environmental implications of modern climate change are stronger thunderstorms and hurricanes, warmer winters, increasing record warm temperatures, weakening of the North Atlantic drift, shifts in rainfall and vector-borne diseases, over-washing of low-lying islands, decreasing ice caps and sea ice, and weakening of the polar jet thereby increasing the frequency of polar vortexes.

According to Sippel and coauthors, since early 2012, global climate change can be detected in every single day's weather! In other words, global climate change is now continuously and instantaneously detected all across the globe every single day. We have, therefore, reached both a climate tipping point and are witnessing a scientific paradigm shift. Sippel and coauthors' work now allows us to point at Australia's 2019–2020 catastrophic fire season and say without doubt that its severity and spatial breadth are the product of global warming. In fact, Van Oldenborgh and coauthors suggest that the Fire Weather Index for NSW and Victoria prior to the 2019–2020 fire season was 30% higher than in 1900, and this was due to anthropogenically induced climate change.

> The Fire Weather Index is a measure of the most intense daily fire risk and monthly severity rating.

Before their work, climate change was framed by climate scientists as a phenomenon that is emerging but could not yet be attributed as a causal mechanism to extreme weather events. This paradigm is dead! We live climate change every single day and will for the rest of our lives.

For instance, the majestic and unique redwood forests of coastal California are in peril. Coastal fogs driven by adjacent ocean upwelling of cool water protect redwoods from summer droughts. According to Johnstone and Dawson, fog reduces redwood sap flow and transpiration, and hence water consumption during dry summers. Between 1951 and 2008, Johnstone and Dawson observed a 33% decline in fog frequency that appears tied to global climate change and the Pacific Decadal Oscillation. This long-term decline in fog frequency has heightened drought sensitivity, and so today the redwoods are dying!

And just to complicate things further, Hye-Mi Kim and coauthors identified in 2009 a fourth ENSO mode, in addition to the typical El Niño, La Niña, and Neutral, that involved warming in the Central Pacific due to climate change that increases the likelihood of tropical cyclone landfall in the eastern Pacific and western North Atlantic. In addition, Tim Barnett and coauthors, in 2005, identified the impact of a warming climate on water availability in snow-dominated regions. They found that global warming triggered rapid snowmelt and spikes in runoff that overwhelm local and regional storage capacities. The consequences of these hydrological changes for future water availability cannot be understated.

2

The UN Intergovernmental Panel on Climate Change (IPCC) report released October 8, 2018, clearly spells out the risks to our planet of continued warming. Since the Industrial Revolution, the earth has warmed 1 °C! Yet the earth will warm another 0.5 °C if we do not take immediate action, and this additional warming will worsen droughts, floods, extreme heat events, and poverty for much of the Global South (IPCC 2018).

All these are difficult enough to contend with, but over such a short geological time span of a century or less, these changes are potentially catastrophic. Without access to inexpensive fossil fuels, adapting to these environmental changes will be much more difficult. Air conditioning could mitigate against warmer temperatures, but not if the electricity becomes too expensive to use it. Areas of the world without access to cheap, abundant air conditioning, which includes many developing and even developed countries (e.g., European countries) or socioeconomically depressed areas in North America, are not likely to adapt to increasing internal air temperature controls under expensive installation and operation forecasts.

Let's be honest—the earth's atmosphere is warming. But it is not due to the Milankovitch cycles, as today the dominant cycle, the eccentricity cycle, is in a cooling phase. Nor is it due to a warming sun or sun-spot activity, given the fact that solar output has declined recently; nor is it due to volcanic activity, as volcanoes emit less than 1% of the total yearly output of CO2 that humans emit into the atmosphere.

The scientific data are, at this point, undeniable; *actual* climate data collected for better part of 40 years and *proxy* climate data gathered from tree rings, ice cores, fossil pollen, ocean sediment, coral, and other historical data clearly illustrate a significant upward trend in temperatures, suggesting the earth has not been this warm in the last 800,000 years. What is more, the impacts of this recent warming are serious and threaten the survival of many hundreds of thousands of species, entire ecosystems, and ways of life. Yes, we are responsible for this warming, but the earth's destiny is still in our hands. John Holdren, Barack Obama's senior advisor on science and technology, has been quoted as saying the following:

> We basically have three choices: mitigation, adaptation, and suffering. We're going to do some of each. The question is what the mix is going to be. The more mitigation we do, the less adaptation will be required and the less suffering there will be.

The problem is that we must, as a global community, act collectively now to mitigate. However, the world's second biggest current GHG polluter and greatest historic contributor to GHGs, the United States, declared on October 24, 2019, that they were pulling out of the Paris climate agreement. Greta Thunberg's plea to the UN Climate Action Summit in 2019 needs to be taken seriously: talking to all countries present, Thunberg stated:

> You are failing us. But the young people are starting to understand your betrayal. The eyes of all future generations are upon you. And if you choose to fail us, I say: We will never forgive you.

> We will not let you get away with this. Right here, right now is where we draw the line. The world is waking up. And change is coming, whether you like it or not.
> Greta Thunberg, September 23, 2019. NPR Transcript of her speech

As part of adapting, we must move away from the fantasy of continued growth. Joseph E. Stiglitz, Nobel laureate of economics, argues that US corporate success, driven by short-term profits with little regard to the impacts of business decisions on the environment or working-class, must change. In 2017, global climate change cost the United States 1.5% of GDP. This will only increase through the twenty-first century, with the agriculture and energy sectors losing hundreds of billions of dollars. Stiglitz makes the point—why should taxpayers pay the enormous economic costs and human costs of suffering when it would be cheaper to mitigate, through technological innovation, employing ideas and policies recommended by the Green New Deal?

James H. Kunstler in 2007, in his book *The Long Emergency,* argues that adapting to climate change in the post-2008 peak oil production environment will be difficult as rising oil prices will reduce funding to develop alternative energy means and climate change mitigation efforts. He suggests we might return to a "large world" where communities will need to be self-sufficient in food, labor, healthcare, and transportation. Kunstler's very pessimistic views are not widely shared. However, urban planners are paying attention and encouraging cities to design communities where work, shopping, entertainment, dining, and living all occur within smaller districts that can be accessed on foot or through public transportation, thereby reducing GHG emissions. Atlanta, Georgia's Atlantic Station with its IKEA, is an upscale business and residential example that was opened in 2005 and built on the former brownfield site of the Atlantic Steel Mill.

2.2 Climate Facts

The potential for climate change was first proposed in the early nineteenth century, but the first definitive articles identifying climate change and CO2 increases were written in the 1960s and 1970s. In 1967, Manabe and Wetherald attempted to model the consequences of doubling global CO2 levels; their prediction and their physically sound model have proved very accurate and robust. In 1976, Keeling and coauthors documented for the first time the recent and very rapid increase in CO2 since WWII. These seminal works triggered deep concern and great angst among the general public, while a paradigm shift among scientists followed, wherein today scientists are increasingly vocal environmentalists.

The Keeling curve is the longest record of human-induced greenhouse gases (GHGs) that are responsible for global warming. The graphic insert in ▫ Fig. 2.2 highlights the seasonal variations in CO2 uptake by terrestrial plants. To clarify, the red curve shows the average monthly concentrations, while the black curve is a statistically smoothed trend.

The problem with CO2 and methane, colloquially known as greenhouse gases (GHGs), is that they absorb long-wave outgoing terrestrial radiation (OTR or

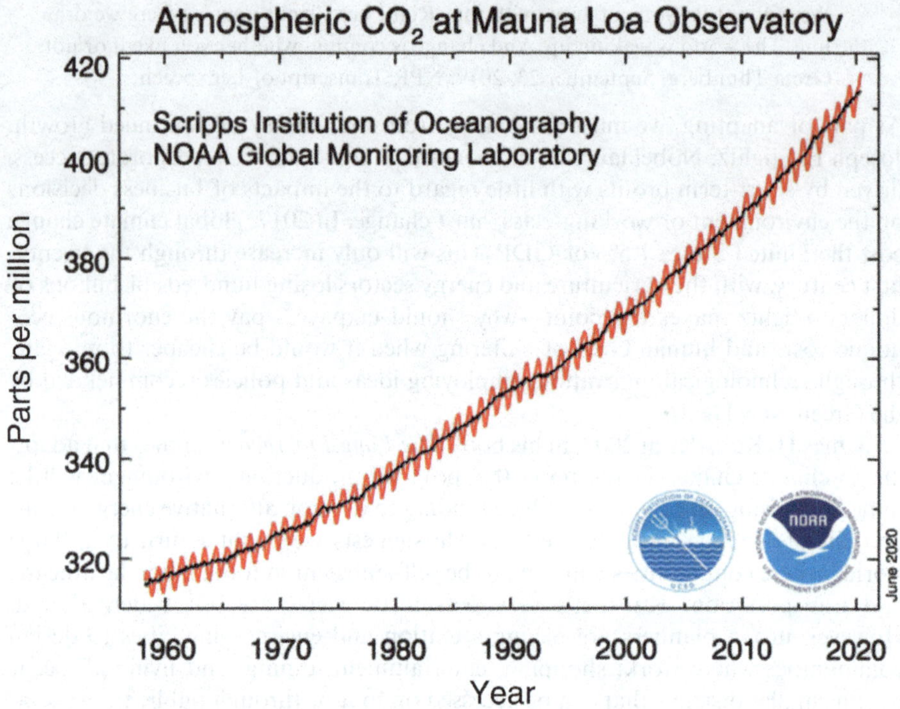

Fig. 2.2 Atmospheric carbon dioxide (CO2) concentrations from 1958 to 2020 directly measured at Mauna Loa, Hawaii (NOAA 2020)

infrared radiation). Both are on the rise globally as an extreme example of non-point-source pollution. OTR is a portion of incoming solar radiation (ISR) that is emitted both by relatively cool terrestrial ground surfaces and by the ocean's surface. Simply put, the earth's surface absorbs ISR, heats up, and releases OTR. To clarify, the surface of the sun is 5778 K or 9941 °F or 5505 °C and emits highly energetic shortwave radiation whose peak output is in the visible light spectrum centered on the color yellow. This is why the sun appears yellow. Earth's atmosphere is mostly opaque to ISR, and ~48% of all ISR that is incident at the top of our atmosphere hits the ground surface. However, we are 92.96 million miles from the sun, so the total ISR at the top of our atmosphere is only 1360 watts per square meter. Although 48% of ISR is incident at the earth's surface and some is reflected due to surface albedo, still an enormous amount is absorbed, heating the land surfaces. As a result, the earth's surface average temperature is 58.62 °F or 14.9 °C as of 2017. As a result, the earth emits low-energy, long-wave (terrestrial) radiation or OTR between 3.0 and 100 μm. Tragically for humans, CO2 absorbs OTR with wavelengths between 4 and 80 μm but not between 8 and 12 μm and heats up our atmosphere. To clarify, if we had no CO2, we would be a much cooler planet. But with too much CO2, we cook!

Methane (CH4) and water vapor also absorb OTR. Water vapor has a similar absorption spectrum as CO2, but methane can absorb 30 times more energy/unit mass than CO_2 because it has a broader OTR absorption spectrum than CO_2. Thankfully, methane is found in small volumes at around 1.8 ppm, but its atmospheric concentration is also growing. In contrast, CO_2 is now over 400 ppm and at the highest level in the last 800,000 years.

Although methane contribution to GHGs is small, according to the US Environmental Protection Agency (EPA), a significant proportion of methane emissions can be traced to landfills and the disposal of food waste. For instance, 1000 kg of biodegradable waste produces between 200 and 400 cubic meters of landfill gas, which is approximately 50–55% methane and 40–45% carbon dioxide (CO_2).

Nitrous oxide also absorbs OTR but in a narrow range of 5–8 μm. NO_2 has increased by 20% since 1750 due to the use of synthetic and manure fertilizers, fossil fuel combustion, and nylon and nitric acid production. However, there is significant spatial variation in NO_2 concentrations around the earth. Between 2001 and 2010, Geddes and his coauthors found, using satellite observations, that the Asia-Pacific exhibited NO_2 of 4.7 ppb, Western Europe 3.9 ppb, North America 3.5 ppb, but Southern America varies between 1.3 and −1.0 ppb, while Africa varies between 0.9 and 0.2 ppb. Rather surprisingly, they found that NO_2 concentrations actually declined over North America by −4.1%/year, over western Europe by −2.5%/year, and over Japan and South Korea by −2.1%/year in the period 2001–2010.

In contrast, ozone (O_3) absorbs incoming solar radiation (ISR) before it enters the lower atmosphere or troposphere (the layer of gases up to 10–14 km above sea level). Naturally occurring O_3 is found in the stratosphere (above the troposphere), where it reacts and absorbs the ISR, releasing heat in the process that traps troposphere gases. Without stratospheric O_3, we would have no atmosphere! This might explain why governments worldwide perceived the depletion of ozone as a very critical threat to humanity and, in a united effort, enacted a global ban of Chlorofluorocarbons (CFCs). Today, O_3 is back to 1970s levels, and the threat posed by O_3 depletion has receded. Nevertheless, less O_3 has meant more ISR incident on the earth's surface, contributing to more global warming.

2.3 Who Is Responsible for GHGs

According to Ritchie and Roser, the biggest anthropogenic contributor to total CO_2 since 1751 is the USA: in fact, the USA has emitted 400 billion tons or 25% of all CO_2 released. This is twice the amount emitted by China over the same time period. Even the EU-28, which has among the strictest emissions legislation and enforcement, has emitted less than the USA, with 22% of all CO_2 since the pre-industrial era. In contrast, rapidly industrializing India and Brazil have not had enough time to contribute much to total global CO_2 emission, and sub-Saharan Africa has contributed ~6% of total CO_2 emissions since the pre-industrial era. Today, China is the single biggest emitter of CO_2 on a national scale, having both the world's largest population (1.4 billion in 2019) and the world's biggest economy. Indeed, according to the World Bank, China emits over 10 billion metric tons of CO_2 annually, while the USA pumps over 5 billion tons of carbon into the atmosphere

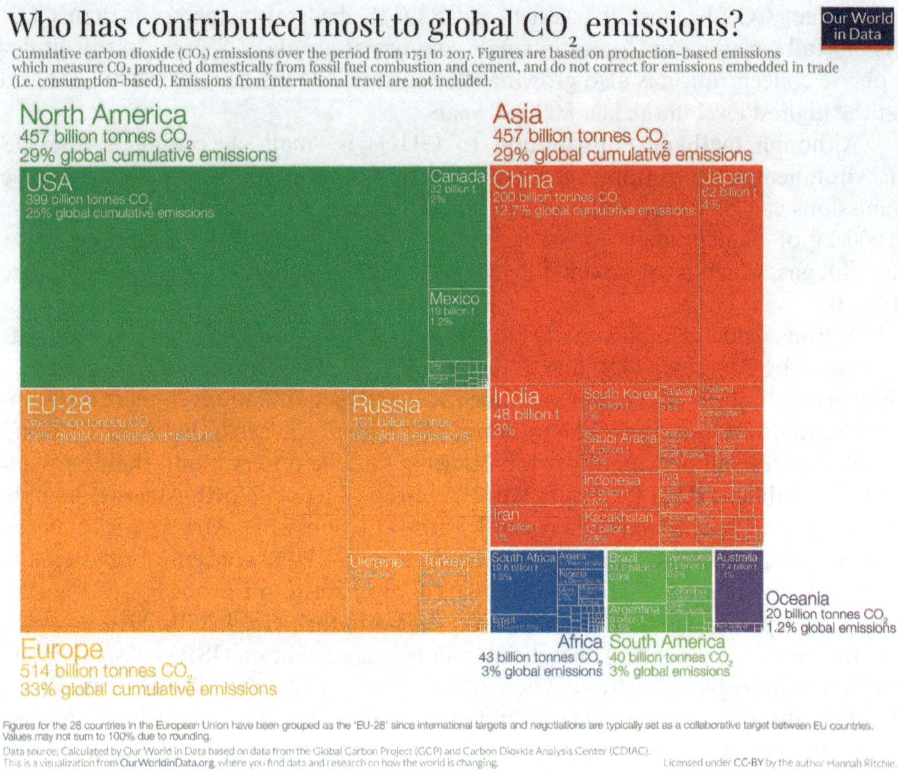

Fig. 2.3 Total CO2 historic contributions by region in 2017. Calculated by Our World in Data based on data from the Global Carbon Project (GCP) and Carbon Dioxide Analysis Center (CDIAC). @Hannah Richie 2017

each year. Yet, China's population is four times the size of the United States', so the average US resident is contributing much more to the global mean CO2 concentration than is the average person in China. Due primarily to its population size, India emits approximately 2.2 billion tons of CO2 each year, while Russia and Japan follow with 1.7 and 1.2 billion metric tons/year, respectively. The rest of the top ten countries in annual CO2 emissions are Germany, Iran, Saudi Arabia, South Korea, and Canada, with total emissions ranging from Germany's 720 billion tons to 537 billion in Canada (Fig. 2.3).

A fairer way to compare a country's contributions to climate change is as per capita (pc) global emissions, since by definition, per capita measures standardize according to population. So, in effect, per capita CO2 emissions allow us to compare countries while accounting for differences in population. Globally, 4.981 metric tons of CO2 is emitted per person (2014). This has increased from 4.194 metric tons in 1990, a 19% increase over nearly 25 years. When we standardize to per capita, oil states in Arabian Peninsula and Caribbean islands yield extremely high pc CO2 measures (20 and higher). This is higher than the USA at 16.5 metric tons/year/pc, which is over three times the global average. In comparison, Germany

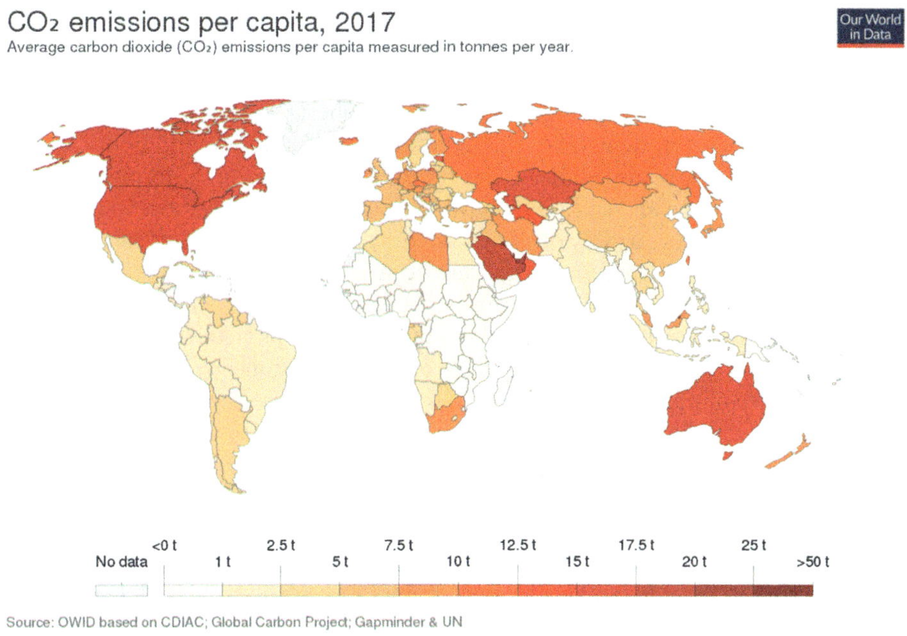

Fig. 2.4 Per capita CO2 in 2017. OWID based on CDIAC; Global Carbon Project; Gapminder & UN

yields 8.9 metric tons/year/pc and China 7.5 which is only about one and a half times the global average. The UK generates 6.5 metric tons/year/pc, Indonesia yields 1.8, and India generates just 1.7 metric tons/year/pc which is over a third of average global per capita emissions (Fig. 2.4).

Today, ~50% of global CO2 emissions are caused by electrical power generation and heat production. Transport and manufacturing industries each contributes ~20%, while residential, commercial, and public services emit ~9%; all other sectors emit 1 to 2% of CO2. Per capita emissions illustrate the massive divide in wealth and technology across the earth. For instance, Qatar, one of the wealthiest countries, is the largest per capita emitter due in large part to desalinating sea water, yet Chad emits at least 1200 times less CO2 per capita.

In contrast, the biggest anthropogenic contributors to total methane (CH4) are mining of coal, oil, and natural gas (20%), enteric fermentation by principally cattle (16%), flooded rice fields (10%), biomass burning (6%), landfills (6%), animal waste (6%), and domestic sewage (4%). This contribution is twice the natural background rates. Although naturally atmospheric hydroxyl radicals break up CH4, these radicals are unable to cope with recent CH4 increases, and as a result methane is increasing at 8% yearly.

Methane is also being released as a by-product of climate change in the Arctic, as thawing permafrost is exposing organic carbon to microbial decomposition. This release of methane is significant and will further warm our atmosphere, with the net effect that more permafrost will thaw releasing more methane in a vicious, runaway negative feedback cycle. Thankfully, there is a fairly limited amount of permafrost in and around the Antarctic; otherwise, the situation would be even more dire. Nevertheless, Schuur and coauthors suggest that the volume of carbon stored in the Arctic permafrost is twice the atmospheric volume of carbon, and a large volume of this is likely to be released in the near future. An additional source of methane might be continental shelf gas hydrate deposits. This is deeply concerning given that Svensen and coauthors have suggested that initial global warming of the Eocene period was possibly due to the melting of marine sediment gas hydrates.

According to Rick Heede and the Climate Accountability Institute, 35% of all carbon emissions since 1965 have been produced by just 20 fuel companies. The vast majority of their emissions were from the combustion of their petrol, jet fuel, natural gas, and coal. Approximately 12% of their emissions are from the extraction, flaring, refining, and delivery of their products. Some are privately owned, while others are national assets; yet, many (both private and public) have received billions of dollars of government subsidies to maintain fuel production and provide subsidized, inexpensive energy sources. The most egregious are Saudi Aramco, Chevron, Gazprom, ExxonMobil, National Iranian Oil Co., BP, Royal Dutch Shell, Coal India, Pemex, and Petroleos de Venezuela.

According to Griffin and Heede, since 1988, in excess of 50% of all GHGs can be traced to just 25 entities. In rank order, they are China (state coal production), Aramco, Gazprom, National Iranian Oil, ExxonMobil, Coal India, Pemex, Russia (state coal production), Shell, China National Petroleum, BP, Chevron, PDVSA, Abu Dhabi National Oil, Poland Coal, Peabody Energy, Sonatrach, Kuwait Petroleum, Total, BHP Billiton, ConocoPhillips, Petrobras, Lukoil, RioTinto, and Nigerian National Petroleum.

Many of these entities have also spent millions fueling climate change denial in order to remain viable. Hiding the real cost of carbon-based energy behind subsidies and denying climate change, and hence hiding the costs of carbon emissions, have meant the maintenance of our oil-based civilization long after we should have been seeking alternatives. Where would we be now if the first electric car developed in 1890–1891 by William Morrison of Des Moines, Iowa, had proved viable!

In addition to subsidizing production of carbon-energy forms, many governments subsidize fuel consumption. In Ecuador, the government ended 40 years of gasoline and diesel subsidies on October 3, 2019. Overnight, diesel doubled in price, while petrol increased by 30%. These subsidies have cost the Ecuadorian government over $60 billion over the last 40 years. In response to the removal of subsidies and subsequent price rises, various indigenous groups including the Confederation of Indigenous Nationalities of Ecuador (CONAIE), students, and transportation unions representing taxi drivers, truckers, and bus drivers went on strike, blocking roads and highways and bringing Ecuador to a halt. Although Ecuador has more sustainable means of energy generation in the form of a substantial HEP program, its railway system was largely abandoned in the 1950s and no new lines have been built since 1908. However, in 2008, President Correa began a program to reactivate the system, though today mostly tourists ride three repaired tracks.

Social media and mobile phone technology are also contributing to GHG emissions. Solnit argues that the rise of transportation network companies can also be linked to increased GHG emissions, although finding digital solutions for package deliveries certainly has reduced transportation times and distances and has contributed to a reduction in GHG emissions. However, companies like Lyft and Uber have stripped customers from public transportation and reduced the number of people who walk or cycle to work. Lyft and Uber offer convenience and comfort but increase GHG emissions, as more people return to vehicular transportation.

2.4 The Scale of Climate Change

The scale of global climate change is hard to comprehend on human timescales. It appears both minor (e.g., what is just one or two degrees of warming?) and overwhelming (e.g., when seven hurricanes were active across the Western Hemisphere in late September 2019). The following may give some sense of proportion to global climate change:

- No one born since 1976 has lived through a colder than average year
- 2012 was among the ten warmest years since record-keeping began in 1880, rising above the long-term average for the 36th year in a row
- 2017 was the warmest year on record
- July 2019 was the hottest month of record
- The period from January through July of 2019 produced a global temperature that was 1.71 °F above the twentieth-century average of 56.9 °F, tying with 2017 as the second-*hottest year* to date on record
- The top 700 meters (about 2300 feet) of world's oceans show warming of more than 0.4 °F since 1969.

We live at a remarkable time in human history. Unlike Greenlanders or Easter Islanders or the Mayan or Anasazi whose political, economic, and social institutions and cultural values failed to prioritize forest and soil conservation and hence their societies collapsed, the vast majority of scientists, environmental managers, farmers, and lay-people do appreciate what we are doing. Yet a small minority of businesspeople, religious leaders (mostly in the USA), and politicians across the world are putting their self-interests above the interest of the rest of the world and the plants and animals with which we share this planet. In his book *Collapse*, Jared Diamond highlighted the insane folly of ignoring climate change today, because although some past societies failed to respond to destructive environmental management (e.g., Greenlanders burning sod [soil] for fuel and Easter Islanders cutting down the last trees), pre-industrial societies in Papua New Guinea, Japan, and Tonga were very well aware of the impact of their deforestation and changed their activities accordingly. In other words, we have no excuse—we have the means to document global climate and environmental change, the means to analyze and plan changes to our negative activities, and the means to implement these positive changes; yet, collectively, the world's leaders are not all on the same page. Yet, anthropogenic climate change is among the most important global issues of our time, environmental or otherwise.

2

There is no Planet B. We have to act now. According to the 2018 Intergovernmental Panel on Climate Change (IPCC) Report, we have just 12 years to act! The IPCC 2018 report is not to be *sneezed at*; it is comprehensive, scientifically valid, robust, and critically important. It was coauthored by 91 climate experts and 133 contributing authors, from some 40 countries, who between them assessed some 30,000 scientific papers. We also cannot ignore people like Greta Thunberg or Autumn Peltier or Sir David Attenborough or Bruno Rodriguez or Jane Goodall or Helena Gualinga or Leonardo DiCaprio or India Logan-Riley or Naomi Klein or Nakabuye Hilda F. or Jamie Henn or Ivy Chipasha or scientists like Jason Box or James Hansen or Charles Keeling (see ▶ Fig. 9.1) or Syukuro Manabe or Jean Robert Petit or Gavin Schmidt, among many, many others who are demanding we change our consumption of resources and our production of GHGs.

We are nearing a tipping point, a threshold, where beyond this threshold, the effects of climate change will be "irreversible" (see Threshold chapter). The International Energy Agency's Greenhouse Gas Research & Development Program estimates the earth's hydrocarbon capacity is ~800 gigatons of CO2 (GtCO2), with the vast majority stored in the oceans. Currently total emissions are 36 GtCO2 annually. As of 2019, we have 21 years left of storage if emissions stay at 36 GtCO2/year. However, between 2017 and 2018, CO2 emissions rose 1.8%. With this in mind, IPCC 2018 hinted that the situation is more critical and the earth's hydrocarbon capacity will be reached in 2030.

Approximately 375 gigatons of CO2 has been absorbed by the earth's oceans since the initiation of the Industrial Revolution in 1705, resulting in a decrease of 0.1 pH or a 30% increase in acidity. The increased concentration of hydrogen ions in seawater reduces the number of carbonate ions in the oceans. According to NOAA, fewer carbonate ions mean calcifying organisms such as oysters, calcareous plankton, and all coral species struggle to maintain their shells. For instance, pteropods, a calcareous plankton, are at the base of the Antarctic Ocean food pyramid. Pteropods are a critically important source of food for juvenile salmon and krill that whales consume. In contrast, photosynthetic algae and seagrasses appear to benefit from elevated ocean CO2 levels.

The rapid acidification of the oceans we are observing today could initiate a marine mass extinction comparable to that that occurred immediately after the Cretaceous Chicxulub impact 66 million years before present. According to Henehan and coauthors, the Chicxulub impact released enormous volumes of SO2, NOx, and CO2 sourced from carbonate rocks valorized by the impact, and additional CO2 from wildfires, biomass decay, and biological weakening of both the terrestrial and marine carbon cycles triggering rapid ocean acidification; in fact, oceans' pH levels did not return to pre-Chicxulub values until 80,000 years after the impact.

2.5 0.5–2 °C above Pre-Industrial Levels: What Is Happening

> Climate-related risks to health, livelihoods, food security, water supply, human security, and economic growth are projected to increase with global warming of 1.5°C and increase further with 2°C (IPCC 2018 report)

According to the most recent estimates from the United Nations' Intergovernmental Panel on Climate Change (IPCC), human-driven (anthropogenic) activities have caused approximately 1.0 °C of global warming since the mid-1800s (i.e., pre-industrial times), with a likely range of 0.8 °C to 1.2 °C. We also know from documented changes in some climate and weather extremes when global warming reached 0.5 °C above pre-industrial levels, that an additional 0.5 °C of warming compared to today's levels would lead to further changes in these extremes (specifically in temperature and precipitation patterns).

Global climatic change driven by increased GHG levels is leading to increased variability, number, and intensity of both hurricanes and tornadoes, and increases in the number of warmer ocean temperature anomalies as El Niño's. Even Goldman Sachs (known for their conservative approach to investments) echo in their 2019 report the doom and gloom reported in the preceding paragraph! They simply state that global climatic change will change disease patterns, intensify heatwave anomalies, generate more destructive weather events such as tornadoes, hurricanes, and even polar vortexes, and negatively impact the quality and quantity of water for agriculture and for human consumption.

Rainfall Global warming's most fundamental effect is to increase global atmospheric moisture, and hence precipitation, as the absolute humidity of air, which is the total mass of water vapor that can occur in g/m^3, increases geometrically as air warms. The absolute humidity (AH) of warm air at 30 °C/86 °F is ~30 g/m^3 of water vapor, at 15 °C/59 °F the AH is ~13 g/m^3 of water vapor, and at 0 °C/32 °F the AH is ~5 g/m^3 of water vapor. In other words, cool air is relatively dry, while warm air is humid. As the earth has warmed, air above the earth's oceans has warmed, and it now holds more water vapor. Kunkel and coauthors found that since 1900, the USA has seen an increasing trend in extreme heavy precipitation events. In this case, a heavy precipitation is a two-day rain total that is exceeded on average only once in a five-year period (▫ Fig. 2.5).

More Intense Heatwave Anomalies The deadly heat anomalies of Chicago and Paris in 2003 will increasingly become the norm. Meehl and Tebaldi suggest that semi-stationary 500 hPa positive height anomalies that cause clear skies, light winds, and intense surficial heat waves are getting stronger, lasting longer, and occurring more frequently because of increasing GHG emissions. These anomalies seem to be atmospherically teleconnected to high precipitation anomalies over India due to intense monsoon events. More intense heatwave anomalies are also expected to affect the

2

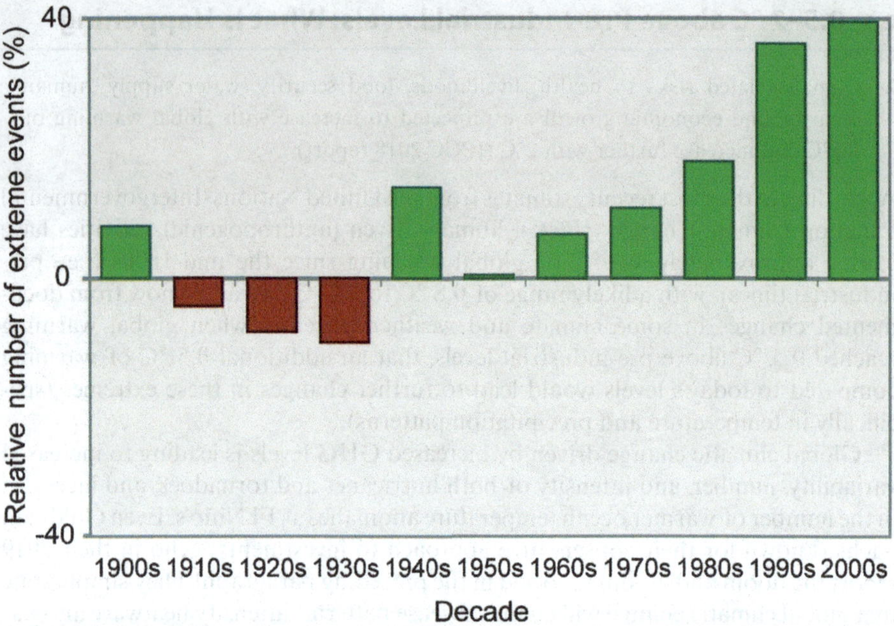

Fig. 2.5 Percentage of heavy two-day precipitation events in the USA between 1901 and 2012 that exceed a typical five-year event. By the 2000s, heavy two-day rainfall events exceeded the five-year return event ~40% of the time. (National Climate Assessment Report 2014: ▶ nca2014.globalchange.gov)

Atlantic Seaboard and northeast of the USA, as well as the Benelux countries and the Balkans.

More El Niños Wang and coauthors have established that since the 1970s, the warm-water anomalies that initiate El Niño events have both grown in magnitude and shifted from the eastern Pacific to the western Pacific as the western Pacific has warmed. This shift in size and location has increased the strength of El Niños and increased their socioeconomic impacts. For instance, the 1997–1998 El Niño was the costliest on record, with some estimates of damage exceeding $96 billion.

Larger More Destructive Hurricanes and Tornadoes According to Webster and coauthors, the variability, number, and intensity of hurricanes have increased since 1850. For instance, their work shows that more categories 1 and 2 hurricanes are becoming categories 3, 4, and 5. Furthermore, Welford and coauthors show that the landfall of Atlantic hurricanes along the Georgia coastline has moved northwards by some 80 km since the 1750s. Knutson and coauthors suggest that anthropogenically enhanced global radiative forcing will lead to 6–34% decrease in hurricane frequency but an increase in hurricane intensity of 2–11% by 2100. According to HURDAT, of the ten most intense Atlantic hurricanes on record (measured by central-eye pressure), only two hurricanes occurred before 1970: Wilma in 1969 and the Labor Day hurricane of 1935 (Fig. 2.6).

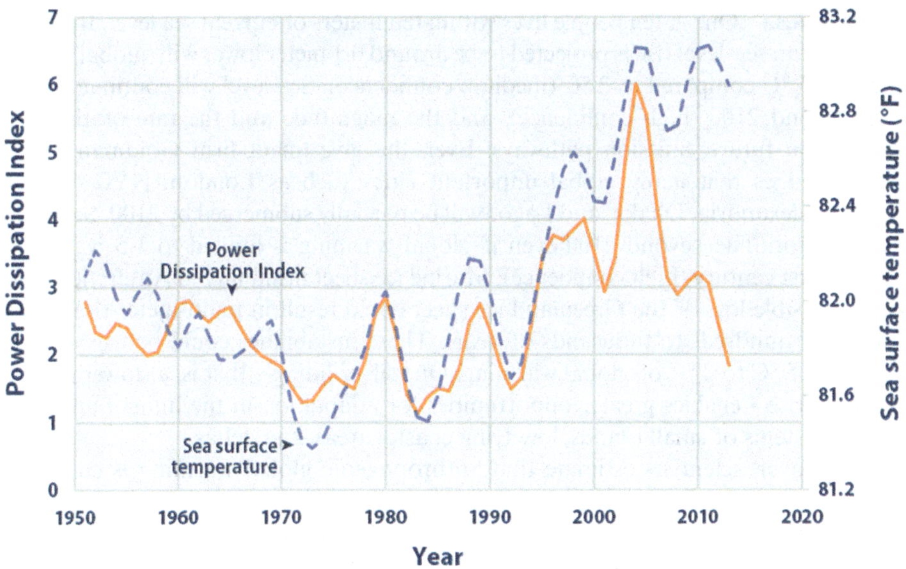

Fig. 2.6 North Atlantic cyclone occurrence by Power Dissipation Index that considers cyclone strength, duration, and frequency. (Data source Emanuel 2007)

Diffenbaugh and Trapp suggest that there will be a net increase in the number of days with high convective available potential energy (CAPE) and strong low-level wind shear. Together, these will increase the number of severe thunderstorms in the USA through the twenty-first century. However, these conditions will vary spatially across the USA, with higher than normal incidence of CAPE in proximity to the Gulf Coast and Eastern Atlantic seaboard. Their projections are being witnessed today as the number, intensity, and deadliness of tornadoes have recently increased in Dixie (Tornado) Alley that ravages East Texas, Louisiana, Mississippi, Alabama, Georgia, and South Carolina from late fall through early spring. The deadly Joplin tornado of May 22, 2011, cost $2.8 billion in damages and killed 158 people in Dixie Alley.

Weakening Polar Jet The polar jet, located along the upper-air boundary between tropical and polar air masses across North America, is weakening as the Arctic warms 2–3 times faster than elsewhere. The slope of the boundary and the velocity of the polar jet are governed by the difference between the temperatures of these two air masses. A fast jet traps polar air, whereas a weak polar jet allows cold air to periodically surge south in the form of a polar vortex. As these polar vortexes penetrate south, they encounter warmer and more humid air than previously, and in response, Cohen and coauthors found these events generate more snowfall and record cold temperatures. In the spring, melting snow contributes to more spring flooding as seen across the Midwest in both 2018 and 2019.

Sea-level Rise One in ten people lives within ten meters of current sea level. By 2100, global mean sea-level rise is projected to be around 0.1 meter lower with global warming of 1.5 °C compared to 2 °C (medium confidence). Sea level will continue to rise well beyond 2100 (high confidence), and the magnitude and the rate of this rise depend on future emission pathways. Even the investment firm Goldman Sachs acknowledges that many global important cities such as London, NYC, Tokyo, Miami, Alexandria, Dhaka, and Lagos will be partially submerged by 2100. Sea-level rise will continue beyond 2100 even if global warming is limited to 1.5 °C in the twenty-first century (high confidence). Marine ice sheet instability in Antarctica and/or irreversible loss of the Greenland ice sheet could result in multi-meter rise in sea level over hundreds to thousands of years. These instabilities could be triggered at around 1.5 °C to 2 °C of global warming. One silver lining—that is, a slower rate of sea-level rise—enables greater opportunities for adaptation in the human and ecological systems of small islands, low-lying coastal areas, and deltas.

Moreover, scientists estimate that anthropogenic global warming is currently increasing at 0.2 °C (likely between 0.1 °C and 0.3 °C) every ten years because of past and current emissions. It is important to note that this human-driven warming will continue for several years, or even decades, after net global greenhouse gas emissions fall to zero, depending upon when we reach that hypothetical net-zero moment. To reiterate, even if human-induced greenhouse gas emissions were to fall to zero, global temperature would continue to increase for several years because of this time lag. Thus, sea-level rise will continue to experience a similar time lag, as we discuss in more detail later in this chapter. As mentioned earlier, populations at disproportionately higher risk of adverse consequences with global warming of 1.5 °C and beyond include socioeconomically disadvantaged populations and local communities dependent on agricultural or coastal livelihoods.

More Carbon, Longer Growing Seasons Wuebbles and Hayhoe argue that the mid-latitudes will experience longer growing seasons as late spring frosts will occur earlier and early fall frosts will occur later. Crop yields will also globally increase as more carbon is emitted into the atmosphere. They estimate that if CO2 reaches 555 ppm (parts per million by volume), wheat yields could increase by 60–100% above early 2000 production levels. However, in the northern hemisphere, the northward migration of insect pests might compromise these yields and reduce profits, as pesticide use would rise.

Future Risks Although there remains some uncertainty concerning the exact magnitude and geographic extent of how these changes to the earth's climate will impact both natural and human-made systems (primarily based upon not knowing exact levels of GHG emissions in the future and the complicated feedback loops involved), scientists have a great deal of confidence that failure to significantly curb or halt global emissions will do irreparable harm to populations and environments worldwide. For instance, scientists around the world have already documented trends in intensity and frequency of some climate and weather extremes over time frames in which approximately 0.5 °C of global warming has occurred.

Future climate-related risks and impacts are highly dependent upon the rate, peak, and duration of warming, such that the IPCC has placed great emphasis on outlining the likely differences between an increase of 1.5 °C above pre-industrial levels and an increase in 2 °C. Recall that we are approximately 1 °C above pre-industrial levels today, and if it continues to increase at the current rate, global warming is likely to reach the 1.5 °C threshold between 2030 and 2052. In addition, the effects of such warming include likely shifts in weather patterns, resulting in some areas receiving flooding rains and other areas experiencing withering droughts. Given the already precarious state of the world food production and safety network, a drought in a prime growing region or in an area in which drinking water is sourced at the surface from lakes and streams could result in serious food and safe water shortages, amongst a host of other problems. Limiting warming to 1.5 °C compared with 2 °C is projected to result in smaller net reductions in yields of maize, rice, wheat, and potentially other staple crops, particularly in sub-Saharan Africa, Southeast Asia, and Central and South America. Furthermore, reductions in projected food availability are larger at 2 °C than at 1.5 °C of global warming in the Sahel, southern Africa, the Mediterranean, central Europe, and the Amazon. Nearly half of all humans already live in water-stressed regions, and these areas will suffer more stress and expand to affect more than 50% of the world's population by 2025. Water-stresses already contribute to a significant movement of environmental migrants, and this will only continue to increase.

Nearly every negative impact on both environmental and human systems increases substantially when models move from the 1.5 °C to the 2 °C scenario. Specifically, climate models project that climate attributes will change immensely across all regions when the scenario shifts from the present-day (1.0 °C above pre-industrial levels) and 1.5 °C range increase, to an increase that surpasses 1.5 °C above pre-industrial levels and marches toward a 2 °C threshold increase. These differences include increases in mean temperature in most land and ocean regions, heat extremes in most inhabited regions, heavy precipitation in some regions, and the likelihood of precipitation declines and/or droughts in other regions. As a result of these extremes, reductions in projected food availability are much larger at 2 °C than at 1.5 °C of global warming in the African Sahel, southern Africa, the Mediterranean, central Europe, and the Amazon. And although it is impossible to know with 100% certainty, climate scientists stress that the 2 °C threshold may lead to long-lasting or even irreversible impacts, such as the permanent loss of some ecosystems and/or ecosystem services. Coral reefs, for example, are projected to decline by a further 70–90% at 1.5 °C, with near total loss (>99%) at 2 °C. Like coral reefs, the risk of irreversible loss of many marine and coastal ecosystems increases with global warming, especially at 2 °C or more. However, limiting global warming to 1.5 °C compared to 2 °C is projected to lower the impacts on terrestrial, freshwater, and coastal ecosystems and to retain more of their services to humans.

2.6 Climate change Vulnerabilities

2

Geographic Contexts Matter Regardless of the future of GHG emissions, the degree of warming, and/or mitigation strategies employed, the simple fact remains that some places will be impacted more than others. For starters, generally warming is greater over land than over water. In addition, some regions experience well above average warming, like the Arctic that is exhibiting warming two to three times the global average. For the most part, regions at disproportionately higher risk include Arctic ecosystems, dryland regions, small island developing states, and developing countries. Indeed, increasing warming amplifies the exposure of small islands, low-lying coastal areas, and deltas to the risks associated with sea-level rise for many human and ecological systems, including increased saltwater intrusion, flooding, and damage to infrastructure. Multiple regional changes in climate also are likely to occur in a global warming scenario up to 1.5 °C compared to pre-industrial levels. These broad, geographically diverse changes include extreme temperatures in many regions; increases in frequency, intensity, and/or amount of heavy precipitation in several regions; and an increase in intensity or frequency of droughts in other regions. Moreover, the implications of these regional climate changes on people's everyday lives are myriad, as this geographic unevenness is slated to take on many forms. For instance, coastal and low-lying areas, impoverished populations, those relying on subsistence agriculture, and those living in the tropics (especially in the southern hemisphere) will remain the most vulnerable. These vulnerabilities encompass issues ranging from negative agricultural impacts, increased frequency of flooding, saltwater intrusion, damage to infrastructure, and threats to ecological and human systems. All of these scenarios become even more likely and more devastating should the level of global warming reach the 2 °C above pre-industrial levels threshold.

The United States Disasters are a composite of social, political, and economic environmental variables extending beyond the natural event that caused it. If the USA, one of the richest nations on earth, continues to exhibit a deeply vulnerable SES-status community, we are in serious trouble. Evidence documented by Webster and coauthors already indicates that high-magnitude weather events such as tornadoes and hurricanes are increasing in size and frequency due to global climate change. If, as seems likely, the role of government in everyday life continues to decline in the USA, risk preparedness will also decline. In a study of tornado risk, Mulilis and Duval found that participants who perceived themselves at a high risk for tornado activity but felt that they could not prepare adequately showed less initiative to prepare for a disaster than those who felt they could adequately prepare; however, this effect only emerged when students felt that they, rather than the government, had the responsibility to prepare. In other words, people must feel responsible, capable, and effective at engaging in such preparation efforts, something deeply vulnerable, politically marginalized SES-status communities across the world struggle to achieve. Typically, women show more concern about disasters than men. Also, people with less than a high school education are far less likely to seek shelter than those with a high school education or greater. In contrast, people with formal education are more likely to take protective measures against future tornado or hurricane activity than people without formal education. Many functionally illiterate people in a lower socioeconomic status (SES-status) often live in mobile homes in the USA and are thus at greater risk of dying in deadly tornadoes or hurricanes. In fact, Sutter and Simmons identified that residents of mobile homes are more than ten times as likely to die from tornadoes than those living in a permanent home.

Southern US states exhibit both physical and social vulnerabilities and/or marginalized populations and high rates of poverty and functional illiteracy. In addition, mobile home occupancy is increasing across rural areas, with concomitant poor communication infrastructure, with populations that do not speak English at home increasing. In particular, US residents living in poverty, which amount to 15.1% of the overall population, have different social and risk event histories that contribute to different risk perception and interpretation. Furthermore, 6% to 23% of people in the United States are functionally illiterate. In fact, the highest prevalence of functional illiteracy subject to tornadoes and hurricanes is in Mississippi, Louisiana, Alabama, and Arkansas. The largest increase in the population of people that do not speak English in their homes (1990–2000) in states with high tornado and hurricane activity is in Georgia, Alabama, Arkansas, and Mississippi (National Center for Education Statistics, 2003). This presents its own risk hazard for many communities, which typically rely on kin and social networks for information, thus frequently delaying risk-aversion behavior. For instance, Mexican Americans tend to accept warnings as credible only after confirming a warning among family and social networks. Finally, southern states may be more economically impacted by severe weather events. For instance, Lazio and coauthors suggest that Alabama exhibits the largest sensitivity (10.1–13.5%) to weather variability as a percentage of gross domestic product by state where tornadoes are common.

Climate Change and Diseases Climate change also increases human, livestock, and arable crop risks and vulnerability to diseases, in particular among humans—tropical diseases such as Zika virus, West Nile virus, Chikungunya virus, malaria, and dengue. Warming temperatures, longer growing seasons, and fewer frosts have meant that dengue and malaria are an increasing risk to humans in Texas and Florida; Spain, Italy, and Greece in Europe; and elsewhere Australia, Chile, Argentina, and South Africa.

The impact of climate change on malaria is hard to predict. Although warmer temperatures in the lowlands are not expected to critically change malaria incidence, malaria is expected to penetrate further up into highland environments both due to increasing temperatures and increases in precipitation. However, changes in land use, increases in population coupled with rapid urbanization, urban-rural migration seeded by tropical forest transition policies of the 1960–1970s in South America, and recent economic development among most Global South countries appear to play a stronger role in the expansion of malaria than in climate change. The built environment simply offers more stagnant pools of water within which mosquitoes can breed. Interestingly, Pinault and Hunter suggest that malaria spread into the highlands of Ecuador between 1900 and 1940 due to a 0.5 °C increase in average daily temperatures and a 1.3 °C warming in minimum daily temperatures, but it was subsequently eliminated through the extensive use of pesticides and deforestation.

Semenza and Suk suggest that in Europe, Lyme disease and tick-borne encephalitis seem to be expanding upslope and northwards due to climate change, as does Zika, dengue, and chikungunya, which are transmitted by Asian tiger mosquitoes, and leishmaniasis, which is transmitted by sandflies. Dengue fever is already expanding its range by responding to increasing vapor pressures or actual humidity due to global warming and warmer ocean surfaces. Increased humidities and increased temperatures both increase mosquito habitats and increase the virus replication. Projections by Simon Hales and coauthors suggest that by 2085, between 50% and 60% of the human population will be at risk of dengue due to climate change, whereas only 30% of the population is currently threatened.

In contrast, the rapid transmission of the Ebola through Sierra Leone, Liberia, and Guinea between the autumn of 2014 and October of 2015 that killed 11,312 people was facilitated by fear of hospitals and the dysfunctional nature of health systems, by high population mobility across state borders in part due to diamond and arms smuggling, by densely populated capitals, and by a lack of trust in authorities after years of armed conflict. Redding and coauthors suggest that in West and Central Africa, continuing animal-human viral spillovers coupled with overburdened and feared healthcare systems will increase the risk of contracting Ebola threefold by 2070. Although near-future climate change will increase the potential ranges of viral hosts, continuing deforestation will likely reduce human-wildlife contacts. However, increased connectivity through increased airline traffic among West and Central African countries and other African countries, Europe, North America, and Asia will increase the likelihood that Ebola will spread beyond Africa repeatedly in the foreseeable future.

In summary, climate change is likely to lead to major population health impacts; however, in Africa, increasing malarial drug resistance, overburdened and feared healthcare systems, and increased poverty likely will have more direct consequences on human health.

2.7 Nonhuman Responses to Climate Change

In terms of biogeography, many animals and plants will run out of space as their ranges will shrink, as they are forced up mountains and mountain ranges. The geographic range of many human and agricultural disease vectors (e.g., insects, rodents) will expand or shrink as climate change continues, possibly resulting in disease outbreaks in formerly disease-free areas. Wildfire seasons are already months longer than 50 years ago and will continue to get longer, while coral reefs have been bleached of their colors.

Among invasive species, earlier springs allow invasive species to push aside native plants. The purple loosestrife, a European invasive, blooms 24 days earlier than it did 100 years ago in New England, while in contrast, the Pennsylvania Bitter Crest only blooms 1 day earlier! This would not be an issue, but purple loosestrife chokes wetlands, killing off cattails among other wetland plants and denying food to dependent species.

Over the last 100 years, Earth has warmed by 0.6 °C, with two main periods of warming between 1910 and 1945 and since 1976. In the last 40 years, organisms have responded in four quite different ways: their phenology (timing of seasonal activities) and physiology (functions of living organisms) has changed; their ranges contracted, shifted, or in some instances expanded; their composition of and interactions within communities changed; and the structure and dynamics of ecosystems significantly changed.

Changes in Phenology and Physiology Among changes in phenology and physiology, earlier leaf unfolding in spring is among the most significant. Today, earlier leaf folding globally generates 3.6 days longer growing season than 50 years ago. Record-keeping among wineries in and around Beaune, France, from 1354 to 2018, is among the longest records of climatically induced biotic change and shows that since 1987, grape harvests occur 13 days earlier than between 1354 and 1987. Grapes are ripening earlier due to the increased summer temperatures.

In the last 20 years, barnacle geese numbers have dramatically increased, as Baltic/North Sea populations bred 6–7 weeks earlier even before peak food availability. Walther, writing in 2002, found that warmer spring weather in Europe disrupted the synchrony between winter moth (*Operophtera brumata*) hatching and oak bud burst, leading to a mismatch between the peak in insect availability and the peak food demands of great tit (*Parus major*) nestlings. But the loss of great tit young in the spring has been offset by increased juvenile survival as well as increased immigration during winter. Thus, the mismatch in timing has not caused a decline in pre-breeding population size.

Weeks and coauthors recently identified that North American migratory bird species are getting smaller and their wings longer across diverse taxa over 40-years as a consequence of global warming. Tarsus, bill length, wing length, and mass of 70,716 individuals across 52 species killed since 1978 by colliding with buildings in Chicago were measured. It appears that migratory bird wing length is increasing as a compensatory adaptation to declining body size and an increased metabolic cost of flight during migration.

Range Change Both vertebrate and invertebrate altitudinal and latitudinal breeding range changes have been observed within the United States, Europe, SE Asia, and Papua New Guinea, while feeding ranges among seabirds in the North Atlantic have also shifted poleward. In South America, even trees have been observed to be on the "move" upslope. But in the long term, all animals and plants that are moving upslope or latitudinally toward the poles will ultimately run out of both space and elevation, consigning them to extinction. In the highland tropics and high Arctic, this reality will occur rather rapidly, while in lowland tropics and mid-latitudes, with more land and elevation to occupy, range-compression extinction events will happen at a slower rate (■ Table 2.1).

2

Table 2.1 Range changes among animals

Species	Data, location	Documented changes	Proposed cause	Source
British birds	Atlas breeding birds 1968–1972, 1988–1991, Britain	18.9 km shift northward in range limits	Climatic warming	Thomas, Lennon (1999)
North American birds	Breeding Bird Surveys 1967–1971, 1998–2002, North America	2.35 km/year northwards	Global climate change	Hitch, Leberg (2007)
Dung beetles in Alps	Discrete beetle surveys between 1993 and 2007, 1982, and 2007 in the Alps	Ranges shifted upward between 44 to 321 meters in elevation	Increasing temperatures	Menéndez et al. (2014)
Vertebrates	Global meta-analysis of vertebrates, Europe, NA, Chile, Malaysia, Marion Island	Vertical 11 m/decade, 16.9 km horizontally/decade	Climatic warming	Chen et al. (2011)
Birds	Ranges from field guides 25 years apart, SE Asia	84 of 485 bird species extended upper elevation limits 399 ± 263 m	Climatic warming, deforestation	Peh (2007)
Andean tree genera	Two discrete surveys, Manu, Peru	2.5–3.5 vertical meters upslope/year	Climatic warming	Feeley et al. (2011)
Balearic shearwater	1980–2003, local, national ornithological record committees, Atlantic	Three times more birds seen in UK and Irish waters	Rising SSTs and shift in plankton, anchovy	Luczak et al. (2011)
48 butterfly species	Finnish National Butterfly Recording Scheme 1992–1996 and 2000–2004	Ranges shifted, on average, 59.9 km northward, 3 species shifted >300 km northward	Warming climate	Pöyry et al. (2009)
Montane birds	Two discrete bird surveys, PNG	Upper-range limits shifted 113–152 m higher, lower-range limits shifted 95–123 m upward	Warming climate	Freeman, Freeman (2014)

▶ **Example**

Among the most critical range-restricted, threatened, and endangered species due to climate change are those species isolated on islands, volcanoes, or isolated mountain ranges. For instance, the ~300 black-breasted puffleg individuals found only on Pichincha Volcano and the Cordillera de Toisán in Ecuador are critically endangered. Their numbers have been declining at >20%/decade for several decades due mostly to agricultural expansion into their range, firewood collecting, and natural and human-induced fires. Although critical habitat is now protected along Yanacocha, the Inca Ditch on Pichincha, and the city of Quito has also adopted the hummingbird as its emblem, climate change coupled with "no place to go" will ultimately cause its extinction. ◀

Elsewhere, the endemic fauna and flora of the Sierra Nevada de Santa Marta, Colombia; Mt. Kilimanjaro and East Usambara Highlands in Tanzania; and Mt. Kenya are all highly vulnerable to climate change.

For those that breed on isolated island mountain ranges and volcanoes but feed elsewhere, such as many seabird species, the threat of climate change is less critical on their breeding sites than is its impact on their food sources. For instance, the Zino's petrel is one of the most endangered seabirds of Europe, but this is due to feral cat predation of eggs and juveniles at their nesting grounds in the highlands of Madeira, Portugal.

In a very interesting twist on climate change adaptations, in 2012, Rachel Pateman discovered that the Brown Argus butterfly caterpillars had changed their diet from eating rockrose plants that grow in chalk hills in southern England to wild geraniums that are more widespread across southern England in response to climate change. Brown Argus butterflies have subsequently expanded northwards ~79 km in 20 years.

Composition of and Interactions Within Communities In southern Switzerland, Walter and coauthors found that vegetation has shifted from indigenous deciduous community to a community dominated by exotic evergreen broad-leaved vegetation. In particular, the shrub layer is today dominated by the growing number of spreading exotic evergreen broad-leaved species that are profiting from milder winter conditions.

Walther and coauthors found substantial impacts on community structure have been observed in coral reefs during periods of warmer than normal sea temperatures. Poised near their upper thermal limits, coral reefs have undergone global mass bleaching events whenever sea temperatures have exceeded long-term summer averages by more than 1.0 °C for several weeks. Since 1979, the incidence of mass coral bleaching is increasing in both frequency and intensity.

▶ **Example**

During the severe 1998 coral bleaching event, an estimated 16% of the world's reef-building corals died, while coral species richness decreased 61% worldwide. Since 2014, severe bleaching of tropical corals has been triggered by persistent record-breaking sea-surface temperatures. The Great Barrier Reef suffered its worst bleaching in 2016, with staghorn and tubular corals suffering a near complete die-off, and is suffering another die-off in 2020. The entire reef suffered in excess of a 27% loss in 2016. ◀

These are akin to a mass extinction event. In fact, coral reefs have suffered the greatest die-offs during the five previous mass extinctions. Irrespective of whether these extinctions were caused by impactors or climate change, all affect the carbon cycle and increase ocean acidity. These extinction events have been so severe in fact that following each mass extinction, coral reefs disappear from the geologic record and do not reappear for typically four million years. Such "coral gaps" would have had catastrophic impacts on tropical fish, as many use coral reefs for nurseries and coral spawn for food. Given that we are witnessing increasing ocean acidity (ocean acidity has increased 0.1 pH since the Industrial Revolution), it is highly likely that we are in the early phase of a sixth marine mass extinction event.

Lindström and coauthors found that across Europe, bird community temperature index (CTI), a composite of individual bird species temperature indexes (STI), lagged behind climate change and that high CTI communities were more successful at their northerly limits than low CTIs over the last 25 years. To clarify, species dependent on warm temperatures have high STI values, whereas bird species dependent on cool/cold temperatures have low STI values. A high CTI value simply means that the bird community has more high value STI species.

However, the picture is far from clear. In California, ranges among both bird and plant species have moved up and down, although among plants, habitat generalists are expanding their ranges upward. In contrast, all frogs are in decline due to exposure to a chytrid fungus.

The Structure and Dynamics of Ecosystems According to Nancy Grimm and coauthors, the Arctic Ocean will be ice-free in late summer by 2050. This could detrimentally impact under-ice phytoplankton blooms that have recently been found to be widespread and contribute significantly to annual primary productivity in the Arctic Sea. Increased thermal stratification due to increased large lakes' and oceans' temperatures is also reducing primary productivity by 1%/year. However, Grimm and coauthors note that increased tree growth at or near latitudinal and altitudinal range limits is seeing trees expand poleward at 20 km/year in the United States. However, tree growth (and hence net primary productivity, NPP) is on decline at the boreal-temperate transition due to climate change–induced drought stress. This decline in boreal NPP seems to mirror declines in rusty blackbirds (>90%), Canada warblers and evening grosbeaks (>75%), boreal chickadees and Olive-sided flycatchers (>70%) according to the Boreal Songbird Initiative. However, the southern Canadian boreal forests have lost 180 million acres due to logging in the last three decades.

In addition, according to Nancy Grimm and coauthors, climate change has increased overwinter insect survival rates and shortened insect development and reproduction cycles, thus increasing the number and size of insect outbreaks. Climate change has also elevated plant drought stress and increased the duration and intensity of the western United States fire seasons and savannah fires in West and East Africa. Climate change has also reduced soil carbon levels and ecosystem sequestering of carbon. Increased droughts across the Western United States over the last 100–150 years have seen native cottonwood-willow riparian forests shrink and be replaced by exotic drought-tolerant tamarisk.

▶ Example

Permafrost is melting, releasing methane, and changing tundra ecosystems. For instance, the Taimyr reindeer herd, the world's largest, has shrunk from a million in 2000 to 600,000 in 2016. Andrey Petrov suggests that climate change coupled with increasing industrialization of the Taimyr Peninsula in northern Russia has triggered this decline. In recent summers, the herd has moved eastward and up in altitude to avoid the increased presence of mosquitoes in the lowlands due to the increased presence of standing water. This has contributed to higher calf mortality as the reindeer have had to migrate further and through more rivers to reach their new summer feeding grounds. Jorgenson and coauthors also found that vegetation succession coupled with the increasing presence of surface water in the high Arctic is thawing permafrost, even though average yearly temperatures remain below 0 °C. ◀

Climate-Induced Extinctions Utilizing big database analyses of online bird range databases including eBird, Global Biodiversity Information Facility, the Avian Knowledge Network, Biodiversity Information Serving Our Nation, the North American Breeding Bird Survey, and many scientific studies, O'Neill and the National Audubon Society used IPCC-projected warming of 1.5 °C, 2.0 °C, and 3.0 °C to identify bird species at risk of extinction in the USA due to global warming. Their analyses suggest that 37% or 389 species are at risk of local or complete extinction, while another 64% are vulnerable to climate change. For instance, utilizing the 3 °C warming projection, the wood thrush would lose 57% of its present breeding range but colonize new areas (to the north) equal to only 20% of its current range. The outlook for the mountain Plover, a range-restricted species of the Great Plains, is even more dire. According to O'Neill, the mountain plover would lose 76% of its current range, but only colonize new areas equal to 1% of its current range. They expect that the mountain plover would become highly vulnerable to extinction.

According to O'Neill and the National Audubon Society, between 87 and 106 bird species scattered across the Boreal Forest, the Midwest, and Northeast of North America would suffer local breeding extinctions if subjected to 3 °C warming. Although these analyses are restricted to North America, these suggest that globally very significant numbers of bird species, and we suspect other animal species as well, will become vulnerable to extinction if we rapidly warm by 3 °C in the near future.

▶ Example

The bushfires in Australia, triggered by intense heating in late 2019 and early 2020, are quite possibly harbingers of the future for other continents. In fact, scientists have warned the Australian government and Australian public for years that global warming would make Australia's summer bushfire season more intense, with more frequent and larger bushfires (see Thresholds for a discussion of tipping points). The Garnaut

Climate Change Review warned that Australia would face catastrophic fire seasons by 2020. Furthermore, Head and coauthors in 2013 suggested that continent-wide warming trends will continue, while rainfall will decrease in temperate areas; together, these will have detrimental impacts on native flora and fauna and agriculture. Conservative estimates from various sources suggest 480 million animals were killed in this latest season including some 8000 koalas in New South Wales: images of a wild thirsty koala seeking water from a human spoke volumes of the problems Australia's fauna and flora were suffering. More than half of the reserves located within the UNESCO listed Blue Mountains have been burnt. This is particularly worrying given the high endemism of these mountains; for instance, the last stronghold of the critical endangered regent honeyeater is here and only 250–400 birds are left! ◄

Australia's summers are hot and dry and prone to bushfires, making the summers very stressful periods, and over the millennia, many Australian animal species have evolved to breed in fall and winter to avoid the stresses of summer. However, this has taken thousands, if not millions, of years of selection, so it is highly unlikely that other animals elsewhere could evolve this response in less than a century to cope with anthropogenic caused warming.

Recently Penn and coauthors reassessed the causality of the single greatest mass extinction that has ever occurred on earth and that terminated the Permian Epoch some 252 million years ago. The end-Permian marine mass extinction killed in excess of 96% of all marine species. They established that a spike in global temperatures caused this extinction event, not a meteorite collision. This is deeply worrying as today's climate change is not ramped, slowly increasing over time, but rather our climate is changing rapidly and would appear like a spike in deaths in a future geologic record.

On the other hand, at the end of the last Ice Age, the earth rapidly warmed, and though not to the degree or speed of today's warming, we can nevertheless, by analogy, compare the two periods. Paleontological analysis of peat bog and lake deposits across North America and Europe suggests that plants and most animals responded slowly and in a lagged manner to the changing temperatures. Trees and shrubs were particularly slow in responding, but this is completely understandable given their lack of mobility and low fecundity. In comparison to today, humans at the end of the Ice Age were having a negligible impact on the environment, although they did contribute to the mass extinction of many animals in excess of 100 kg. However, they had not altered much of the physical landscape such that plants and trees were able to recolonize land once overrun by glaciers. Today, much of the world's landscape is not available for recolonization. Beetles, on the other hand, responded rapidly, their mobility, high fecundity, and willingness to utilize even the most degraded habitats proving valuable assets.

Perhaps one of humanity's greatest fears will come true—humble, meek insects such as beetles and cockroaches will inherit the earth!

2.8 Global Limits to Growth, a Warning to Humanity

As stated at the start of this chapter, climate change is not an esoteric "them-not-us" issue: it is a fundamental game-changer for the modern standard of living that the Western world has become accustomed to over the last century of development. But we have known for a long time the devastating consequences of a "business-as-usual-approach" to environmental management.

In 1972, a report to the Club of Rome titled "The Limits to Growth" that among academics triggered a reevaluation of the value of globalization, Meadows and coauthors utilizing general systems theory and computer modeling proposed 12 scenarios for population growth, industrial production, natural resource consumption, and pollution production for the globe for the next 100 years. Although optimistic, *Limits to Growth* (LTG) stressed that the earth faced a grave challenge—the world must support technological, cultural, and institutional change in order to avoid exceeding earth's environmental carrying capacity. All LTG scenarios projected that the earth would reach its physical limits of natural resource depletion and its capacity to absorb emissions from industry and agriculture in the twenty-first century. Limits to population and industrial growth in LTG were not abrupt; rather, LTG suggested, as the world population continued to expand, more capital would be needed to cope with pollution, thereby reducing the amount of capital for industrial support and development. At some point, industrial output levels could not be supported. LTG argued that as industrial output slumped, so would food output and other forms of consumption and population growth would cease and/or collapse. But before collapse, the social and economic inequality gap across the globe would continue to grow.

Hirsch, in a critic of globalization, argued that an increasing proportion of benefits people derive from consumption is due to status effect. In other words, those with status benefit, while those without status do not, thereby increasing the divide at a global level among those peoples in the Global North and those peoples in the Global South. So, although LTG suggests societal collapse will occur in the twenty-first century, many societies (and individuals), especially in the Global South, are already trapped in unremitting poverty and debt, struggle to feed themselves, are denied land ownership, jobs, education, and healthcare, and hence are in the spasms of society collapse. It should not be surprising then that the world is in the grip of a migration crisis from the Global South to the Global North, triggered by environmental change and societal problems. If industrial societal collapse did occur, the outlook for the majority of the world's population according to LTG would be grim. Collapse would be accompanied by failing health, conflict, and ecological devastation. However, De Ponti and coauthors found that organic farming yields are typically 80% of conventional farm yields, suggesting that industrial inputs into agriculture could be substantially reduced and thereby offer a means to reduce agricultural pollution.

In 1992, a 21-year reevaluation of LTG by Meadows and her coauthors found that humanity had already exceeded earth's environmental carrying capacity. *Beyond the Limits* (BTL) argued that in order to sustain humans on earth, govern-

2

ments and individuals must reassess their own life goals, develop more sustainable technologies, and limit the impacts of full globalization on the earth. Concurrent with reassessments of LTG, in 1992, the Union of Concerned Scientists published "World Scientists' Warning to Humanity"! The core of their message was that human activities including GHG production and global warming, unregulated marine fish exploitation, ozone depletion, groundwater depletion, soil depletion, deforestation, the destruction of global ecosystem services, and continuing human population growth were critically threatening global life support systems across our planet. The authors declared that *we must* wean ourselves off fossils fuels as soon as possible; *we must* develop more sustainable, smaller, cheaper-scale energy production systems for use by Global South countries; *we must* stop deforestation and loss of agricultural land; *we must* improve human welfare and expand conservation and recycling; *we must* improve social and economic conditions of all peoples and eliminate poverty; and *we must* reduce all manner of inequalities especially gender across the world. Sadly, all but ozone depletion has been ignored.

In 2002, a 30-year reevaluation of LTG by Meadows and coauthors found some refreshing changes. New sustainable energy technologies were being developed and adopted. For instance, according to Domonoske, in March 2018, Portugal produced 103.6% of its mainland electrical energy needs from renewable energy. Population growth rates declined between 1972 and 2002, while wealth increased, and point-source pollution declined as environment laws were enacted and enforced; yet, CO_2 pollution increased. Although total caloric consumption grew in both the Global North and Global South, per capita grain production peaked in the mid-1980s, and oceanic fish production declined as fish stocks crashed due to overfishing, with 90% of all large marine predators disappearing. Nevertheless, the authors concluded the world is still in overshoot mode and major technological, cultural, and institutional change is still needed if LTG predictions are to be avoided.

Although the 30-year reevaluation of LTG was generally positive, Turner suggested that the collapse of the global economic system still occurs in the mid-twenty-first century if the world follows a business-as-usual approach identified by LTG. Elsewhere, the International Geosphere-Biosphere Program (IGBP) and Steffen identify the "Great Acceleration" in socioeconomic trends that capture consumption and pollution, and earth system trends in sensitive gases (e.g., CO2, CH4, NO, O3), surface temperatures, ocean acidification, fishing, and land-use change continue through 2010 irrespective of the world economic downturn of 2007. IGBP found that only ozone is declining and is expected to reach 1980 levels before the mid-2000s. Here critics of LTG point to the successful worldwide implementation of the Montreal Protocol and suggest similar protocols could save the global environment and allow business-as-usual to proceed. What these models fail to illustrate, however, is the catastrophic impact these trends are having now and will have in the future on the global environment, its ecosystems, and its species. These broad-based models also fail to capture how dependent humans are on ecological services (here defined as any process arising from healthy ecosystems, such as purification of water and air, pollination of plants, and decomposition of waste). For example, LTG, BTL, and/or IGBP also fail to illustrate the role globalization and technology have on poaching or any form of hunting, or how sprawl or deforestation or agricultural intensification impacts species survival and ecosystem services.

In 2017, the authors of the "Warning to Humanity" suggested we have failed in all our efforts or lack thereof except ozone depletion. Instead, *we have* accelerated the sixth mass extinction begun in the 1600s (see Extinction chapter); *we have* increased our consumption of meat and increased associated GHG emissions; *we have* increased deforestation; *we have* increased agricultural land losses, and *we have* accelerated climate change. On the 40th anniversary of the world's first climate conference held in Geneva in 1979, Ripple and coauthors of the "Warning to Humanity" reiterated their mantra—*we must* end population growth, *we must* end deforestation, *we must* stop mining fossil fuels, and *we must* stop eating meat. However, Ripple and coauthors did identify several positive trends, fossil fuel asset disinvestment is rising, fossil fuel prices are rising, and human fertility rates are declining. And, just as Stiglitz has advocated, *we must* reduce excessive consumption of resources by affluent and wealthy individuals, classes, and countries. Yet, despite some promising signs, Ripple and coauthors note that climate change has reached an irreversible Thresholds (see Thresholds chapter) which, if crossed, would cause significant, irreparable damage to society, to all ecosystems and the services they provide, and all economies.

2.9 Climate Change Denial and Other Assorted Environmental Repudiations?

Is it a fear of change itself, and a radical change in lifestyle that is necessary to combat climate change, that deeply worries people? The gut reactions in social media to Alexandria Ocasio-Cortez's "Green New Deal" have been extreme and yet illustrate a deep-seated contradiction. Many, many lay people fear that traditional food options will be eliminated such as beef steaks, pork chops, and fried chicken. Yet they love pigs, cows, and chickens. Here is the contradiction—they love domestic animals, but cannot or will not connect the industrial farming of these animals and their inhumane treatment to the meat they love and eat, nor can they, or will they, change their consumption patterns to ease the inhumane conditions these animals suffer or the GHGs they produce.

Is it the fear of big government that is necessary to legislate change that stops many, many people from accepting the need to react to climate change? Cotgrove and Duff suggest environmentalists have to continue to challenge the central values and ideology of our industrial society. The "business-as-usual" approach does not challenge us to adapt, change, our ideology or our lifestyle.

On the other hand, the cataclysmic rendering of climate change has created a so-called climate despair, where climate change is viewed as an unstoppable force that will lead to human extinction. This very bleak apocalyptic vision has convinced many people that we cannot help the earth and all that is left is to be raptured! Their view is "why bother" to recycle or conserve or drive fuel-efficient vehicles or plant trees or listen to those amongst us who are concerned about the environment. Those who adhere to a "climate despair" ideology are, in reality, the most irrational, the most radical, not the environmentalists trying to save the earth, its people, and creatures.

Nevertheless, large numbers of people across the world, from teenagers in Sweden, Ecuador, and Argentina, to pop-stars and film-stars, to politicians and beyond, realize actions to halt climate change are necessary, and yet these actions are not being addressed.

In the United States today, Alexandria Ocasio-Cortez, Democratic representative for New York's 14th congressional district, is the new face of this challenge, and as such will continue to enrage right-wing anti-environmental Republican politicians, their supporters, and US evangelicals. Yet following the first Earth Day in 1970, protecting the environment was a bipartisan issue in the USA, both Presidents Nixon and Ford enacted significant environmental policies—Nixon created the Environmental Protection Agency, while Ford signed the Toxic Substances Control Act. But this bipartisanship did not last. President Reagan's conservative administration was staunchly anti-environmental, and his rhetoric and actions have become the cornerstone of successive Republican administrations. As a result of this political climate, the USA is openly hostile to environmentalists of any sort—be they activists or enthusiasts. A demonstration outside of the Republican National Congress illustrates the problem—a Young Republican sponsored group was observed demonstrating against birding, by suggesting birders just watch birds to observe bird sex! Of course, this idea is ludicrous, but clearly the Young Republicans were trying to reduce, marginalize, or otherwise damage a group of people, who, though small in number, do care about the environment.

Today, President Trump is bitterly opposed to any environmental legislation, and he is supported by the Republican Party and most US evangelicals. President Trump is undoing as much environmental legislation as possible by revoking previous environmental legislation and putting anti-environmental people into positions of power within all federal environmental agencies. The cult of personality created around Reagan by Republicans since Reagan's death explains much of this dynamic. In contrast, today's evangelicals enthusiastically embrace both the "end-of-times" and "rapture" as a means to escape the "corrupted" earth. So rather than trying to conserve or preserve the earth and its plants and animals, they are just interested in saving themselves and/or exploiting the earth and its plants and animals, the opposite of Leopold's Land Ethic or Singer's *Animal Liberation* or Næss' Deep Ecology.

The rise in populist nationalism across the world since 2010 led by Donald Trump in the USA, Brexiteers in the UK, Marie Le Pen in France, the Five-Star movement in Italy, Jair Bolsonaro in Brazil, Putin in Russia, and Tony Abbott in Australia funded in part by ultra-rich "aristocrats," represents a significant and alarming challenge to environmentalists. Salil Benegal from DePauw University found that 84% of white Americans that exhibited the highest level of racial resentment dispute anthropogenic climate change.

Brooke Harrington argues that nationalism has been sold as "a war for the little guy." Trump's rallying call to "Drain the Swamp" in Washington, DC, is very much part of this scheme, as is Brexiteers slogan "Take Back Control." Harrington argues that although the ultra-rich and ultra-nationalists seem strange bedfellows,

Fig. 2.7 Coauthor Mark Welford talking to villagers in the Eastern Highlands of PNG about their perception of local and global climatic change and ecotourism. 2015@Theresa Welford

they want the same thing: "both seek to weaken or dismantle international alliances that constrain them." This means dismantling or weakening the EU, the UN, all international environmental legislation, and refusing to sign new environmental legislation. The George H.W Bush and Trump administrations both failed to sign the 1992 Kyoto Protocol and 2015 Paris Agreement, and both used a prisoner's dilemma argument stating that if they sacrifice in this direction while others do not, the USA would no longer be competitive. Both administrations argued that the benefits of carbon reduction are experienced by all countries but must be paid by individual countries, so they fear countries freeloading while others invest time and tax dollars or restrain their behavior to save the *commons*. By contrast, Ronald Reagan did sign the Montreal Protocol of 1987 to eliminate CFCs due to the ozone crisis, even though the elimination of CFCs was very costly.

Sadly, climate denial is not just an American or European thing. In 2019/2020, Australia's ruling party has been vehemently anti-climate change yet is suffering its worst bushfire season.

▶ Example

On the other hand, rural Papuans (see Fig. 2.7), among some of the most remote peoples on earth (yet most green), understand and are worried by climate change. They fear for their community, their children, and their livelihoods as they experience first-hand the impacts that climate change is causing them. They are also very well aware that it is the global, rich North that is responsible for the changes they are observing and dealing with. Yet they also wanted to work with foreigners from the Global North through ecotourism to diversify their income streams. The community is well aware of the commodification of birds-of-paradise by birders and was keen to discuss this with one coauthor of this book. They have two pairs of blue birds-of-paradise on their community lands above their village. Incredibly, they were also worried and conflicted about ecotourism contributing to global CO2 emissions! ◀

2.10 What Can Governments Do? What Can We Do?

As individuals, companies, charities, and government organizations, especially those entities in the Global North, we can drive less and fly less, use more public transport, can ride-share, and use virtual digital conferences to mitigate our carbon fuel usage. We can reduce military use of carbon fuels by running more digitally simulated wars. We can reduce our consumption of meat. We can recycle more plastics and find alternatives to plastics such as bioplastics made from avocado seeds. We can employ more renewable forms of energy production—wind, tide, or solar. We can fund more fusion research. We can encourage municipalities to generate electricity through biofuels.

Sadly, climate change will impact those individuals and countries with less wealth and technology. The 2006 Stern Review on the Economics of Climate Change suggests that external damages inflicted by climate change on the world's economies will cost between $10–$350/ton of carbon emitted. Simply put, poor nations with little financial wriggle room will lose more than rich nations. Mitigation of GHGs by improving energy efficiency in transportation, energy generation, and food production; using renewable forms of energy production; protecting soil; and preventing deforestation and reforesting to lessen future impacts of climate change is expensive and difficult to implement, particularly among Global South nations that have limited government revenues. Yet Global South countries are amongst the lowest contributors to global GHG. For instance, Kiri Hanks of Oxfam established that the UK produces more GHGs in the first three weeks of January than the combined total of the following nine Africa countries: Burkina Faso, Cameroon, Ethiopia, Guinea, Madagascar, Malawi, Nigeria, Tanzania, and Uganda.

❯ Attempts to reign in and curb GHG emission by the Global North have been hit-and-miss. Yet, since 1750, those nations that react to and innovate first to energy transitions are those nations that are successful in the post-energy transition era.

▶ Example

For instance, the United States dominated the sperm whale fuel oil economy in the mid-nineteenth century. In addition to using it as a heating fuel, sperm whale oil was used for lubrication and for the manufacture of soap and candle wax. However, after sperm whale whaling peaked in 1846, sperm whale whalers were forced to catch smaller sperm whales in colder waters, utilizing larger, more costly boats that could stay at sea longer to catch and process the more elusive whales. This drove the price of sperm whale up to $1.77/gallon in 1856. This, in part, triggered investment in refining innovations in kerosene distillation from crude oil funded by the likes of John D. Rockefeller that, by 1895, saw petroleum costs drop to 7 cents/gallon. This energy transition occurred in a period of laissez-faire economic policy where political protectionism of businesses was not considered or enacted. ◀

The *1992 UN Framework Convention on Climate Change* (UNFCCC or otherwise known as the Convention) proposed to reduce emissions to 1990 levels by 2000. Although their voluntary, nation-by-nation approach has all but failed, nations in the Global South, unfettered by multinational corporation lobbying, have increasingly demanded binding climate change agreements. Since Rio, the Convention has been ratified by 197 countries. Among several successful actions initiated by the Convention is industrialized nations' funding of climate change in developing countries through a system of grants and loans managed by the Global Environment Facility. The Convention also obtained an agreement among the 197 signatories whereby sustainable technologies developed in industrialized countries would be shared among developing nations. In addition, the Convention forced all industrialized countries to regularly report their climate change policies and measures.

The *1997 Kyoto Protocol* ratified in 2005 demanded that its 127 signatory nations reduce GHG emission to levels below those of 1990. In what became a standard response, the United States did not ratify the protocol, arguing that limiting industrialized countries disadvantaged them economically compared to those industrializing nations that were not capped. The successor to Kyoto—the *2009 Copenhagen Conference*—tried to get industrialized nations to pay and package climate change mitigation actions enacted among industrializing nations and reward those countries that reduced deforestation. Sadly, nothing legally binding was committed to and no GHG targets set.

In 2010, the *UNFCCC Cancun Agreements* were signed and established the Green Climate Fund, the Technology Mechanism, the Cancun Adaption Framework, Fast-start Financing, and Forest Management Reference Levels to support climate change mitigation and adaptation activities in developing countries. These actions represented the most comprehensive package agreed upon to help developing nations tackle climate change. Furthermore, the Cancun Agreements established among all 197 signatory nations, excluding the United States, the goal of keeping global average temperatures below a 2 ° C rise (above pre-industrial levels) this century.

In 2015 the *Paris Agreement* went further than Cancun and demanded signatories limit global temperature rise below a 1.5 °C threshold this century. The Paris Agreement also sought to expedite financial and technology flows to developing countries and enhance global climate change action transparency. Limiting global warming to 1.5 °C above pre-industrial levels is projected to reduce risks to marine biodiversity, fisheries, and ecosystems, and their functions and services to humans, as illustrated by recent changes to Arctic Sea ice- and warm-water coral reef ecosystems. Pathways limiting global warming to 1.5 °C with no or limited overshoot would require rapid and far-reaching transitions in energy, land, urban and infrastructure (including transport and buildings), and industrial systems sustainable water management.

From Rio to Cancun, the US government did not sign nor participate in any carbon mitigation. However, the US federal government did sign the Paris Agreement in 2015. However, President Trump formally announced that the USA would retract itself from the Paris Agreement obligations by 2021. In response, many city mayors and administrations across the USA stated they would continue to cooperate with UNFCCC even if the federal government does not!

During the *2019 Madrid conference*, known as COP25, the USA, Brazil, Japan, Australia, and Saudi Arabia created gridlock in a successful attempt to maintain a business-as-usual approach and block stronger global GHG emission rules and a more robust carbon-trading system. Alden Meyer of the Union of Concerned Scientists responded to this deliberate gridlock with the following criticism:

> » I've been attending these climate negotiations since they first started in 1991, but never have I seen the almost total disconnection we've seen here... in Madrid between what the science requires and the people of the world demand, and what the climate negotiators are delivering. Rebecca Leber, Mother Jones, December 14, 2019.

In the post-COP25 environment, climate change has become an us-versus-them scenario! The rich, Global North (except the EU) wants to maintain a business-as-usual approach in terms of resource consumption and pollution production while forcing the Global South to absorb the most damaging aspects of climate change. This is so arrogant, so racist, so neocolonial in nature, it is hard to accept that these are the policies of democratically elected governments. Instead, Schneider suggests it feels like the Global North has returned to the era of the US robber baron industrialists of the nineteenth century such as Carnegie, J.P. Morgan, Rockefeller, and Vanderbilt, among others, and former European colonial trading companies such as the East India Company and the Congo Free State that monopolized trade, "engaged in unethical business practices, exploited workers and paid little heed to their customers or competition."

In contrast, across Europe, taxing GHG emitting vehicles is one means of addressing climate change. The EU charges value-added tax (VAT) of 20% on most fuel and 5% on domestic heating fuels; these are known as fuel taxes. Vehicle excise duties (VED) are also charged, and in the United Kingdom, these two taxes generated GB£32 billion in 2009 in tax revenue. VED taxes are prorated on car fuel efficiency. For those cars registered on or after April 2017 (as of Oct 2019), a sample of the first year's tax is as follows: emissions of less than 50 CO2g/km are charged GB£50 or less. Thereafter, there is a graduated rising cost, so by 101–110 CO2g/km, owners are charged at least GB£140 and as much as GB£170, for over 255 CO2g/km, owners are charged GB£2125–2135. Rates for the second year's VED drop to a maximum of GB£152.25. Electric cars are not subject to VED, and of course, owners do not pay any fuel taxes. These taxes are all aimed at minimizing GHG emissions and have resulted in UK citizens purchasing fairly small fuel-efficient cars. In Norway, 58% of all new car sales in 2019 were electric vehicles.

Furthermore, hydroelectric power (HEP) contributes nearly all Norway's electrical needs, making electric car usage truly green and sustainable; furthermore, electric cars are exempt from all sales, importation, and road taxes. And the Norwegian government has funded public electric charging stations across the country.

Another means to reduce GHG emissions and congestion is to implement congestion charges. In the UK, congestion charges are in operation in London and Durham. In most Italian towns, commercial buses are subject to congestion charges; for example, in Venice, these are in excess of 300 euros/day. Similarly, New York City instituted its Metropolitan Commuter Transportation Mobility Tax (MCTMT) in 2019.

Climate Change Adaptations Theoretically, a wide range of adaptation options are available to reduce the risks to natural and managed ecosystems, including ecosystem-based adaptation, ecosystem restoration, and avoid degradation and deforestation, biodiversity management, sustainable aquaculture, and drawing on local and indigenous knowledge in the utilization of natural resources. However, the capacity for and likelihood of climate-resilient development differs between and within regions and nations, as myriad development contexts and vulnerability types exist. The principles of social justice and equity must serve as the foundation for decision-making within and across communities at multiple geographic scales for impoverished, underrepresented, and/or disadvantaged communities to avoid a worst-case scenario. Simply put, without a commitment to social justice, disadvantaged people and places will suffer disproportionately the negative impacts of climate change. This commitment must be central at every level of policy/decision-making, from local municipalities to global agreements, in order to sustain (much less improve) rural livelihoods. Considerations of those most vulnerable because of socioeconomic status, racial/ethnic/indigenous identity, and/or multiple other aspects of identity must remain at the forefront when there are discussions and considerations regarding rural livelihoods across the globe (e.g., efficient irrigation, social safety nets, disaster risk management, and community-based adaptation). Similarly, there must be a focus on relatively disadvantaged urban residents in resiliency planning for cities of all sizes, whether discussions center on green infrastructure and sustainable land use/planning or low-carbon emissions transportation options.

Several coastal regions in the Global North and Global South have begun to address the risks of sea-level rise through altering coastlines by way of building sea walls or jetties; these techniques are known collectively as shoreline hardening. Indeed, shoreline hardening exists in approximately 15% of US coastlines, primarily in more populated, above-average socioeconomic status coastal areas. Research has demonstrated, however, that such approaches to climate adaptation potentially impact local ecologies negatively through habitat alteration and augmenting erosion.

Carbon Dioxide Removal and Geo-Engineering According to recent IPCC reports, renewables will need to supply 75–80% of energy needs globally by 2050 if we are to avoid a scenario of greater than a 1.5 °C increase from pre-industrial levels. Moreover, if fossil fuels are to continue to source electricity generation, they will need to include carbon dioxide capture and storage alongside increased use of nuclear as a share of energy sources. Most modeled scenarios necessitate substituting coal with other energy sources if we are to reach the year 2050 with less than 1.5 °C increase from pre-industrial levels.

There are a number of potential technological and/or engineering carbon dioxide removal (CDR) measures, including afforestation and reforestation, land restoration and soil carbon sequestration, bioengineering and carbon capture and storage (BECCS), and direct air carbon capture and storage (DACCS). However, very few of these are mature and ready to contribute (i.e., effectively scalable and affordable) to significantly alter the amount of CO2 in the atmosphere today. This does not mean we should not continue to conduct research and development in these areas, especially given the current state of scientific knowledge and pace of technological advances. We absolutely should. Nevertheless, the urgency of the climate crisis indicates that the global community should not reasonably expect that a multitude of such approaches will be available and affordable on a large scale any time soon. Thus, counting on such technologies to save us from the impending climate crisis is little more than a pipe dream.

At this juncture, therefore, limited resources would be better utilized by implementing adaptation strategies focused on reducing vulnerabilities of human and natural systems in concert with sustainable development. Global South countries, for instance, might work to address current and future vulnerabilities related to food, water, disaster risk, poverty, and inequality into economic development and urban planning (as some currently are, e.g., Cartagena, Colombia, and some cities in Bangladesh). Such a comprehensive approach to sustainable development would work to ensure that the communities and geographies most vulnerable to climate change are supported in efforts to adapt to the future's most likely scenarios. Infrastructural improvements can play an integral role in adaptation strategies; as stated by the IPCC (2018), "Increasing investment in physical and social infrastructure is a key enabling condition to enhance the resilience and the adaptive capacities of societies." This might mean directing resources toward flood mitigation in low-lying areas (including, but not limited to, coastal regions), attention to agricultural challenges in areas dependent upon subsistence agriculture, or planning for increased and intensified tropical storm activity in coastal regions, to name but a few examples.

Chapter Summary

Returning to John Holdren's comment, we have three choices: mitigation, adaption, or suffering. We will suffer unless governments, corporations, and individuals employ all their efforts in mitigation and adaptation rather than playing a wait-and-see or business-as-usual approach. Sadly, those Global North countries and China most responsible for total GHG emissions will suffer the least, while those least responsible will suffer the most due to the inherent vulnerabilities of poor countries. However, the Global North will see enormous climate change-induced migrations.

According to the Environmental Justice Foundation, 21.5 million people were uprooted between 2008 and 2016 by climate and weather-induced disasters. Although the least developed countries account for less than 1% of all GHG emissions, they account for 99% of all deaths from climate- and weather-induced disasters. It is worth noting that the US military, in contrast to President Trump's administration, considers climate change among the most dangerous "threat multipliers" likely to spark local, regional, and global political instability. The Arab Spring, the war in Syria, and the Boko Haram terrorist insurgency have all been linked to impacts stemming from climate change.

References

Barnett, T. P., Adam, J. C., & Lettenmaier, D. P. (2005). Potential impacts of a warming climate on water availability in snow-dominated regions. *Nature, 438*, 303–309.

Benegal, S. D. (2018). The spillover of race and racial attitudes into public opinion about climate change. *Environmental Politics, 27*(4), 733–756.

Boreal songbird initiative: Keep the forest singing. https://www.borealbirds.org/threats-boreal-birds. Accessed 29 Oct 2019.

Chen, I. C., Hill, J. K., Ohlemüller, R., Roy, D. B., & Thomas, C. D. (2011). Rapid range shifts of species associated with high levels of climate warming. *Science, 333*(6045), 1024–1026.

Cohen, J., Pfeiffer, K., & Francis, J. A. (2018). Warm Arctic episodes linked with increased frequency of extreme winter weather in the United States. *Nature Communications, 9*(1), 869.

De Ponti, T., Rijk, B., & Van Ittersum, M. K. (2012). The crop yield gap between organic and conventional agriculture. *Agricultural Systems, 108*, 1–9.

Diffenbaugh, N. S., Scherer, M., & Trapp, R. J. (2013). Robust increases in severe thunderstorm environments in response to greenhouse forcing. *PNAS, 110*(41), 16361–16366.

Elsner, J. B., Liu, K.-B., & Kocher, B. (2000). Spatial variation in major U.S. hurricane activity: Statistics and a physical mechanism. *Journal of Climate, 13*, 2293–2305.

Environmental Justice Foundation. (2019). *Climate change & conflict*. https://ejfoundation.org/what-we-do/climate/climate-change-conflict

Feeley, K. J., Silman, M. R., Bush, M. B., Farfan, W., Cabrera, K. G., Malhi, Y., et al. (2011). Upslope migration of Andean trees. *Journal of Biogeography, 38*(4), 783–791.

Freeman, B. G., & Freeman, A. M. C. (2014). Rapid upslope shifts in New Guinean birds illustrate strong distributional responses of tropical montane species to global warming. *Proceedings of the National Academy of Sciences, 111*(12), 4490–4494.

Garnaut, R. (2008). *The Garnaut climate change review*. Cambridge: Cambridge.

Geddes, J. A., Martin, R. V., Boys, B. L., & van Donkelaar, A. (2015). Long-term trends worldwide in ambient NO2 concentrations inferred from satellite observations. *Environmental Health Perspectives, 124*(3), 281–289.

Griffin, P., & Heede, C. R. (2017). The Carbon Majors Database. *CDP Carbon Majors Report 2017*, p. 14. Climate Accountability Institute.

Grimm, N. B., Chapin, F. S., III, Bierwagen, B., Gonzalez, P., Groffman, P. M., Luo, Y., Melton, F., Nadelhoffer, K., Pairis, A., Raymond, P. A., & Schimel, J. (2013). The impacts of climate change on ecosystem structure and function. *Frontiers in Ecology and the Environment, 11*(9), 474–482.

Gubler, D. J. (1998). Resurgent vector-borne diseases as a global health problem. *Emerging Infectious Diseases, 4*, 3. Special Issue.

Hanks, K. (2019). *Carbon footprint comparisons*. Oxfam. https://oxfam.app.box.com/s/qx9npbxljbw-md94uz2uwx049nwwigc8l

Harvey, C. A., et al. (2018). Climate change impacts and adaptation among smallholder farmers in Central America. *Agriculture & Food Security, 7*, 57.

Head, L., Adams, M., McGregor, H. V., & Toole, S. (2014). Climate change and Australia. *Wiley Interdisciplinary Reviews: Climate Change, 5*(2), 175–197.

Henehan, M. J., Ridgwell, A., Thomas, E., Zhang, S., Alegret, L., Schmidt, D. N., Rae, J. W. B., Witts, J. W., Landman, N. H., Greene, S. E., Huber, B. T., Super, J. R., Planavsky, N. J., & Hull, P. M. (2019, October). Rapid ocean acidification and protracted Earth system recovery followed the end-Cretaceous Chicxulub impact. Proceedings of the National Academy of Sciences, 201905989. https://doi.org/10.1073/pnas.1905989116.

Hitch, A. T., & Leberg, P. L. (2007). Breeding distributions of North American bird species moving north as a result of climate change. *Conservation Biology, 21*(2), 534–539.

Hu, Z.-Z., Yang, S., & Wu, R. (2003). Long-term climate variations in China and global warming signals. *Journal of Geophysical Research, 108*(D19), 4614. https://doi.org/10.1029/2003JD003651.

IPCC. (2018). Global warming of 1.5 °C: an IPCC special report on the impacts of global warming of 1.5 °C above pre-industrial levels and related global greenhouse gas emission pathways, in the context of strengthening the global response to the threat of climate change, sustainable development, and efforts to eradicate poverty. IPCC SR1.5.

IPCC. (2014). Climate change 2014: Mitigation of climate change. Contribution of Working Group III to the fifth assessment report of the Intergovernmental Panel on Climate Change [Edenhofer, O., R. Pichs-Madruga, Y. Sokona, E. Farahani, S. Kadner, K. Seyboth, A. Adler, I. Baum, S. Brunner, P. Eickemeier, B. Kriemann, J. Savolainen, S. Schlömer, C. von Stechow, T. Zwickel and J.C. Minx (eds.)]. Cambridge University Press, Cambridge and New York.

Johnstone, J. A., & Dawson, T. E. (2010). Climatic context and ecological implications of summer fog decline in the coast redwood region. *Proceedings of the National Academy of Sciences, 107*(10), 4533–4538.

Jorgenson, M. T., Romanovsky, V., Harden, J., Shur, Y., O'Donnell, J., Schuur, E. A., Kanevskiy, M., & Marchenko, S. (2010). Resilience and vulnerability of permafrost to climate change. *Canadian Journal of Forest Research, 40*(7), 1219–1236.

Keller, D. P., Lenton, A., Littleton, E. W., et al. (2018). The effects of carbon dioxide removal on the carbon cycle. *Curr Clim Change Rep, 4*, 250. https://doi.org/10.1007/s40641-018-0104-3.

Kim, H.-M., Webster, P. J., & Curry, J. A. (2009). Impact of shifting patterns of Pacific Ocean warming on North Atlantic tropical cyclones. *Science, 325*(5936), 77–80.

Knutson, T. R., McBride, J. L., Chan, J., Emanuel, K., Holland, G., Landsea, C., Held, I., Kossin, J. P., Srivastava, A. K., & Sugi, M. (2010, February 21). Tropical cyclones and climate change. *Nature Geoscience*. https://doi.org/10.1038/ngeo779.

Kunkel, K. E., Karl, T. R., Easterling, D. R., Redmond, K., Young, J., Yin, X., & Hennon, P. (2013). Probable maximum precipitation and climate change. *Geophysical Research Letters, 40*(7), 1402–1408.

Kunstler, J. H. (2007). *The long emergency: Surviving the end of oil, climate change, and other converging catastrophes of the twenty-first century*. New York: Open Road+ Grove/Atlantic.

Labbé, T., Pfister, C., Brönnimann, S., Rousseau, D., Franke, J., & Bois, B. (2019). The longest homogeneous series of grape harvest dates, Beaune 1354–2018, and its significance for the understanding of past and present climate. *Climate of the Past, 15*(4), 1485–1501.

Lazo, J. K., Lawson, A., Larsen, P. H., & Waldman, D. M. (2011). U.S. economic sensitivity to weather variability. *Bulletin of the American Meteorological Society, 92*(6), 709–720.

Lindström, Å., Green, M., Paulson, G., Smith, H. G., & Devictor, V. (2013). Rapid changes in bird community composition at multiple temporal and spatial scales in response to recent climate change. *Ecography, 36*(3), 313–322.

Luczak, C., Beaugrand, G., Jaffre, M., & Lenoir, S. (2011). Climate change impact on Balearic shearwater through a trophic cascade. *Biology Letters, 7*(5), 702–705.

Meadows, D. H., Meadows, D. H., Randers, J., & Behrens III, W. W. (1972). The limits to growth: A report to the club of Rome. USA: Potomac Associates

Meadows, D. H., Meadows, D. L., & Randers, J. (1992). *Beyond the limits: Global collapse or a sustainable future*. London: Earthscan Publications Ltd.

Meadows, D., & Randers, J. (2012). *The limits to growth: The 30-year update*. USA: Routledge.

Menéndez, R., González-Megías, A., Jay-Robert, P., & Marquéz-Ferrando, R. (2014). Climate change and elevational range shifts: Evidence from dung beetles in two E uropean mountain ranges. *Global Ecology and Biogeography, 23*(6), 646–657.

Morelle, R. (2016). *World's largest reindeer herd plummets*. BBC News: Science & Environment. https://www.bbc.com/news/science-environment-38297464. Accessed 30 Oct 2019.

Mulilis, J. P., & Duval, T. S. (1997). The PrE model of coping and tornado preparedness: Moderating effects of responsibility. *Journal of Applied Social Psychology, 27*(19), 1750–1766.

National Center for Education Statistics. (2003). http://nces.ed.gov/pubsearch/pubsinfo.asp?pubid=2003067

New, M., Liverman, D., Schroder, H., & Anderson, K. (2011). Four degrees and beyond: The potential for a global temperature increase of four degrees and its implications. *Philosophical Transactions of the Royal Society A, 369*, 6–19.

Nicholas, T. (1999). Clogs to clogs in three generations? Explaining entrepreneurial performance in Britain since 1850. *The Journal of Economic History, 59*(3), 688–713.

NOAA. (2013). *Ocean acidification*. https://www.noaa.gov/education/resource-collections/ocean-coasts-education-resources/ocean-acidification. Retrieved 9 Oct 2019.

NOAA. (2020). *Assessing the global climate in January 2020*. https://www.ncei.noaa.gov/news/global-climate-202001

NSIDC Notes. (2010) Issue 73. nsidc.org

Ocasio-Cortez, A. (2019, February 7). H. Res. 109 Recognizing the duty of the Federal Government to create a Green New Deal. https://ocasio-cortez.house.gov/gnd

O'Neill, D. (2019). Survival by Degrees: 389 bird species on the brink. Audubon. NAS. Audubon's Climate Science Series Reports.

Pateman, R. M., Hill, J. K., Roy, D. B., Fox, R., & Thomas, C. D. (2012). Temperature-dependent alterations in host use drive rapid range expansion in a butterfly. *Science, 336*(6084), 1028–1030.

Penn, J. L., Deutsch, C., Payne, J. L., & Sperling, E. A. (2018). Temperature-dependent hypoxia explains biogeography and severity of end-Permian marine mass extinction. *Science, 362*(6419), eaat1327.

Pinault, L. L., & Hunter, F. F. (2012). Malaria in highlands of Ecuador since 1900. *Emerging Infectious Diseases, 18*(4), 615.

Post, J. D. (1977). *The last great subsistence crisis in the Western World*. Baltimore: Johns Hopkins University Press.

Pöyry, J., Luoto, M., Heikkinen, R. K., Kuussaari, M., & Saarinen, K. (2009). Species traits explain recent range shifts of Finnish butterflies. *Global Change Biology, 15*(3), 732–743.

Redding, D. W., Atkinson, P. M., Cunningham, A. A., Iacono, G. L., Moses, L. M., Wood, J. L., & Jones, K. E. (2019). Impacts of environmental and socio-economic factors on emergence and epidemic potential of Ebola in Africa. *Nature Communications, 10*(1), 1–11.

Rifkin, J. (2013). *The European dream: How Europe's vision of the future is quietly eclipsing the American dream*. London: Wiley.

Ripple, W. J., Wolf, C., Newsome, T. M., Galetti, M., Alamgir, M., Crist, E., Mahmoud, M. I., Laurance, W. F., & 15,364 Scientist Signatories from 184 Countries. (2017). World scientists' warning to humanity: A second notice. *BioScience, 67*(12), 1026–1028.

Ripple, W. J., Wolf, C., Newsome, T. M., Barnard, P., & Moomaw, W. R. (2019). World Scientists' warning of a climate emergency. *BioScience*, biz088, 10.1093/biosci/biz088. Published 5 Nov 2019.

Ritchie, H., & Roser, M. (2017). CO_2 and greenhouse gas emissions. *Our World in Data*. https://www.OurWorldData.org

Schneider, S. (2019). *Robber Barron, United States history*. Encyclopedia Britannica.

Schuur, E. A., Bockheim, J., Canadell, J. G., Euskirchen, E., Field, C. B., Goryachkin, S. V., Hagemann, S., Kuhry, P., Lafleur, P. M., Lee, H., & Mazhitova, G. (2008). Vulnerability of permafrost carbon to climate change: Implications for the global carbon cycle. *BioScience, 58*(8), 701–714.

Shepherd, A., Ivins, E., Rignot, E., & Fettweis, X. (2019). Mass balance of the Greenland Ice Sheet from 1992 to 2018. *Nature, 579*, 233–239.

Shirer, W. L. (1990). *20th century journey: A native's return, 1945–1988* (Vol. 3). NYC: Simon and Schuster.

2

Sippel, S., Meinshausen, N., Fischer, E. M., Székely, E., & Knutti, R. (2020). Climate change now detectable from any single day of weather at global scale. *Nature Climate Change, 10*(1), 35–41.

Svensen, H., Planke, S., Malthe-Sørenssen, A., Jamtveit, B., Myklebust, R., Eidem, T. R., & Rey, S. S. (2004). Release of methane from a volcanic basin as a mechanism for initial Eocene global warming. *Nature, 429*(6991), 542.

Solnit, R. (2019). In its insatiable pursuit of power, Silicon Valley is fueling the climate crisis. *Guardian Thursday*, 10 Oct 2019. https://www.theguardian.com/commentisfree/2019/oct/10/power-silicon-valley-climate-crisis-big-tech-profitable?CMP=Share_iOSApp_Other

Steffen, W., Broadgate, W., Deutsch, L., Gaffney, O., & Ludwig, C. (2015). The trajectory of the Anthropocene: The great acceleration. *The Anthropocene Review, 2*(1), 81–98.

Stiglitz, J. M. *Corporate greed is accelerating climate change. But we can still head off disaster*. CNN Business Perspective, Sun April 21, 2019. https://www.cnn.com/2019/04/21/perspectives/joseph-stiglitz-earth-day-economy/index.html

Sutter, D., & Simmons, K. M. (2010). Tornado fatalities and mobile homes in the United States. *Natural Hazards, 53*, 125–137.

Thomas, C. D., & Lennon, J. J. (1999). Birds extend their ranges northwards. *Nature, 399*(6733), 213.

Trapp, R. J., Diffenbaugh, N. S., Brooks, H. E., Baldwin, M. E., Robinson, E. D., & Pal, J. S. (2007). Changes in severe thunderstorms during the 21st century caused by anthropogenically enhanced global radiative forcing. *PNAS, 104*(50), 19719–19723.

Van Hoof, T. B., Bunnik, F. P., Waucomont, J. G., Kürschner, W. M., & Visscher, H. (2006). Forest re-growth on medieval farmland after the Black Death pandemic—Implications for atmospheric CO2 levels. *Palaeogeography, Palaeoclimatology, Palaeoecology, 237*(2–4), 396–409.

Van Oldenborgh, G. J., Krikken, F., Lewis, S., Leach, N. J., Lehner, F., Saunders, K. R., van Weele, M., Haustein, K., Li, S., Wallom, D., Sparrow, S., Arrighi, J., Singh, R. P., van Aalst, M. K., Philip, S. Y., Vautard, R., & Otto, F. E. L. (2020). *Attribution of Australian bushfire risk to anthropogenic climate change*. World Weather Attribution. https://www.worldweatherattribution.org/bushfires-in-australia-2019-2020/.

Velcogna, I., Mohajerani, Y., Geruo, A., Landerer, F., Mouginot, J., Noel, B., Rignot, E., Sutterly, T., van den Broeke, M., van Wessem, J. M., & Wiese, D. (2020). Continuity of ice sheet mass loss in Greenland and Antarctica from the GRACE and GRACE Follow-On missions. *Geophysical Research Letters*. https://doi.org/10.1029/2020GL087291.

Wang, B., Luo, X., Yang, Y. M., Sun, W., Cane, M. A., Cai, W., Yeh, S. W., & Liu, J. (2019). Historical change of El Niño properties sheds light on future changes of extreme El Niño. *Proceedings of the National Academy of Sciences, 116*(45), 201911130.

Webster, P. J., Holland, G. J., Curry, J. A., & Chang, H.-R. (2005). Changes in tropical cyclone number, duration, and intensity in a warming environment. *Science, 309*(5742), 1844–1846.

Weeks, B. C., Willard, D. E., Ellis, A. A., Witynski, M. L., Hennen, M., & Winger, B. M. (2019). Shared morphological consequences of global warming in North American migratory birds. *bioRxiv*, 610329.

Welford, M. R., Bossak, B. H., & Gibney, E. J. (2017). Archival evidence of secular changes in Georgia hurricanes: 1750–2012. In *Hurricanes and climate change* (pp. 35–54). Cham: Springer.

Wuebbles, D. J., & Hayhoe, K. (2004). Climate change projections for the United States Midwest. *Mitigation and Adaptation Strategies for Global Change, 9*(4), 335–363.

Zenghelis, D. (2006). *Stern review: The economics of climate change*. London: HM Treasury.

3

Extinctions

Contents

Learning Goals

After reading this chapter, you will be able to:

- Analyze historical trends and geographic patterns of extinctions.
- Explain how islands extinctions move to continents.
- Evaluate deforestation on continents that has led to isolated islands of varyingly degraded habitats.

3.1 Extinction Is Forever a Game-Changer

In this chapter, we discuss the human-environmental processes that are leading to the sixth mass extinction.

> Because of human-caused extinctions and climate change, the *Anthropocene* was recently defined as the period in earth history where human activity has irrevocably changed global climate and the environment, and many refer to the *Anthropocene* as the sixth mass extinction.

But we should not limit our association of the *Anthropocene* with a nonhuman mass die-off; rather, we should also note that the pollution in the *Anthropocene* has caused unnecessary and early deaths of thousands, if not millions, of humans.

> We therefore recommend the *Anthropocene* be known as an *omnicide* (we mentioned this in the first chapter), where the current global anthropogenic ecological catastrophe will likely extinguish most other species on this planet.

Whether or not you call this part of earth history the *Anthropocene,* and hence an *omnicide*, the ultimate cause of all our global environment problems is the greed inherent at the heart of capitalism. Capitalism has driven ocean-island extinctions, near-continental island extinctions, and now continental extinctions since the 1600s. Even before 1600, the Roman Empire's enormous greed for animal blood sports held in the coliseums decimated populations of North African cheetahs, lions, leopards, and elephants. Patel and Moore prefer to refer to the post-1600s as the Capitalocene, and it does not pay its bills. In other words, the environmental costs of capitalism are rarely, if ever, paid. And you simply cannot buy back an extinction! Jurassic Park is still a fairy tale as of 2020.

Simply put, extinctions serve as a filtering process; extinction episodes result in decreases in diversity and a simplification of biotic communities, which themselves are then more prone to chaotic fluctuations in community structure and species numbers. Extinction can be the direct result of human activity (e.g., hunting), or typically indirectly through deforestation, habitat destruction (e.g., plowing up prairies, construction of built environments), and habitat alteration. Irrespective of the human activity, those prone to extinction include species with restricted geographic ranges, species-poor families, species that are dietary or habitat specialists, species of large body size (and are edible), species that exist at the upper trophic level (carnivores), species with low population densities, and species with poor dispersal abilities.

Until recently, these extinctions were limited to islands; yet, sadly, extinctions have now arrived on continents as human populations have rapidly grown, enormous areas of natural habitat have been converted to agricultural land, and hunting for food and poaching for "fake" medicines and jewelry have increased. Moreover, since WWII, the introduction of alien species has expanded, resource consumption and pollution have grown, and all of these trends have exploded since the 1990s. The process of global integration has been active for a couple of thousand years, starting with Alexander the Great, then the Silk Route, and broadened by British, French, and Spanish colonialism to include much of the world. Yet, it was the rise of Japan and, more recently, newly industrializing economies (NIEs) of Asia, including India and China, the fall of the Soviet Union, and the demise of the Council of Mutual Economic Assistance (CMEA) that accelerated globalization in the 1990s. By 1993, western Europe's, North America's (including Mexico), Japan's, and NIEs' share of world merchandise exports peaked at more than 80% of world trade. Thereafter, trade continued to increase, and more countries were fully integrated into this vast system. At the same time, across the world, we have seen a rise in national populism and income inequality, both driven by the world's elite economic aristocracy. Brooke Harrington argues that nationalism, which attempts to serve the "disenfranchised majority," in fact, helps the elites because both nationalists and aristocrats are anarchists seeking to harness and/or degrade the power of national governments and supranational organizations such as the EU, the UN, and any and all global environmental legislation. Poaching for elephant ivory, rhino horn, pangolin scales, and helmeted hornbills' casques, among many other items, has hit all-time highs since the 1990s. In addition, the ease of transporting either illegal drugs or illegal wildlife trade has been facilitated by the vast increase in container traffic, since in 1990–1999, the global container shipping network expanded at four times the rate of global GDP growth.

To compound the pressures wildlife struggle with, both the USA and the UK, in 2019, once more allowed the importation of animal trophies, in particular, heads and bodies of, among others, lions and giraffes. This relaxation of decades-old restriction caters to the world's elite economic aristocracies that support, lobby, and otherwise pander to such nationalistic governments. Japan also resumed commercial hunting of whales in 2019. And now the USA is building a US-Mexico border-wall that will isolate the few jaguars north of the border (most are in Arizona) from source-populations in Mexico, which will likely condemn the US jaguars to a localized extinction.

3.2 Biodiversity Crises

Today, approximately 24% of 4600 mammal species and approximately 11% of 10,250 bird species are threatened, which means these species are likely to become endangered in the near future. And if they do become endangered, they are highly likely to become extinct. Yet extinction is a natural process! Recent estimates suggest that most species evolve, survive, and go extinct over a million-year cycle. And with approximately 10 million species thought to inhabit earth, this converts to 10 extinctions every year. However, over the last 100 years, a time recently classified as the *Anthropocene*, Pimm and coauthors argue that the extinction rate is 10–100 times the geologic rate. By 1993, just mammal and bird extinctions alone were in the order of 1–3 per year! Given past, current, and future carbon emissions, once oceans absorb 310 gigatons of CO_2—and forecasts from the Intergovernmental Panel on Climate Change suggest by 2100 we will have achieved this threshold—the sixth mass extinction of life will begin (Rothman 2017). At least one previous mass extinction killed in excess of 96% of all life. So, the future outlook is grim indeed, but what of the past extinctions, and are there any spatial and temporal patterns and processes underlying this extinction crisis? Can we look at the past both to explain local, regional, or continental-wide extinctions and to gain insight into trying to reduce the likelihood of a future global mass extinction event?

Unlike any other previous entity, save the evolution of plants and their production of O_2, the evolution of *Homo sapiens* between 250,000 and 350,000 years before present (BP) and our migration out of Africa about 60,000 years BP have changed the globe irrevocably. We have recently changed the global climate, while in the past and present, our technological innovations and our consumption of natural resources have allowed humans to alter soils, landscape, hydrology while poisoning and/or killing entire species across the globe. So, although extinctions are natural processes, humans have initiated waves of extinctions that have hit different places and different animals at different times. For instance, it is estimated that the Hawaii islands have lost over 28 bird species, 72 taxa of snails, 74 taxa of insects, and 97 taxa of plants since the arrival of Captain James Cook in 1778. But even before Cook's arrival, native Hawaiians exterminated a number of Hawaiian honeycreepers. The arrival of early humans on different continents also triggered Late Pleistocene megafauna extinctions. More recently, according to Thompson, at a global scale, whaling triggered a near-mass extinction of oil-rich whales. But the crisis in whale-oil production in the post-sperm whale capture peak of 1846 saved whales and forced our carbon-based civilization to transfer to the consumption of petroleum. In 1846, there were 640 US whaling ships, some three times the size of all the other whaling nations put together. These ships represented the fifth largest producer of GDP in the USA; by 1900, the number and value had dropped 90%, but petroleum production had soared.

These extinctions appear to mirror or lag behind Kondratiev cycles. Daniel Šmihula identified six modern long-wave economic and societal cycles, each the

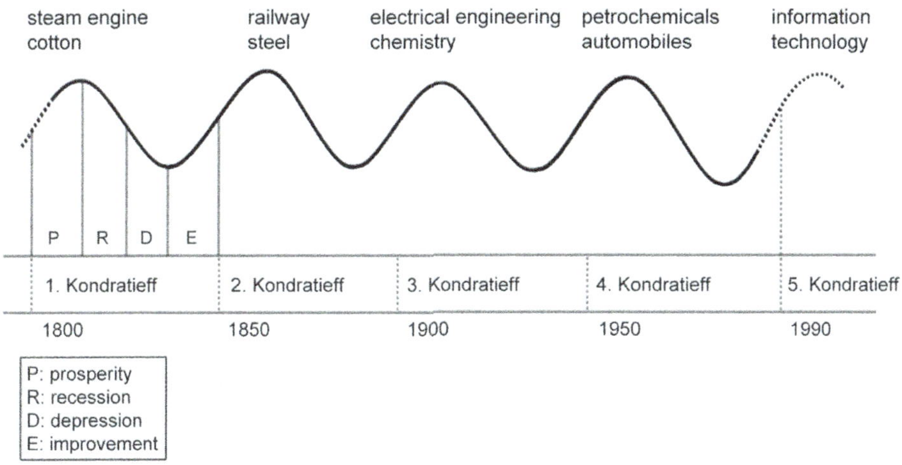

Fig. 3.1 Kondratiev waves: A schematic drawing showing the "World Economy," over time, according to the Kondratiev theory

product of a specific technological revolution (Fig. 3.1: just five are shown, the first was from 1600–1740), and six pre-modern technological waves beginning in 1900–1100 BCE. For instance, four innovations let the USA dominate whaling in the 1800s; these were, first, bigger, better ships supplied with good maps, second, new harpoon technology, third, innovation in winch technology, and, lastly, the use of percentage bonuses amongst all whaling crew to increase whale captures.

The peak in whaling and whale oil consumption occurred in the 1840s, and subsequent declines in whale oil production contributed to the stagnation of the global economy. However, the collapse of whale oil production triggered technological innovation in the refining of other oil sources, and in subsequent exploration, mining, and utilization of crude oil, with the first modern oil well drilled near Baku, in modern day Azerbaijan, in 1848.

3.3 Pre-1600 Extinctions

But before the advent of modern human civilizations in Mesopotamia 6000–7000 years BP, and shortly after the end of the Pleistocene glacial period, 35 genera (55 sp.) of mammal, most of these over 100 kg, went extinct in North America. This is twice as many extinctions as in previous glacial periods. In the Pleistocene, massive megafauna extinctions are unknown before the earliest arrival of humans or occur shortly after a revolution in human technology, such as the development of Acheulean hunting cultures in Africa and the Clovis culture that evolved in-and-around 11,000 years BP in North America. In contrast, in Australia, megafauna extinctions begin 20,000–25,000 years after humans arrived approximately 65,000 years BP.

But all occur at a time of massive ecological change precipitated by Pleistocene climatic changes, suggesting that human hunting (see Threshold ▶ Chap. 4) may have delivered a final *coup de grace* to animals struggling to cope with ecological, climate, and meteorite-impact stresses and has been referred to as an *associated blitzkrieg overkill* event.

3

Pre-1600 AD island extinctions, established by recent archeological evidence, suggest that the arrival of early humans to islands, and not just continents, had catastrophic impacts. In the Caribbean, the newly arrived indigenous peoples contributed to the extermination of many animals including giant sloths, giant rats, and particularly those bird species that have developed flightlessness. Remains of the flightless extinct Cuban giant owl were abundant in cave deposits up to 11,700 years ago. In contrast, the Cuban macaw survived until the late 1800s. Similarly, the arrival of indigenous peoples to Madagascar and New Zealand triggered ratite extinctions. Ratites are large, flightless birds—for example, elephant birds on Madagascar and the nine species of moa in New Zealand. Incredibly, the moa survived Māori hunting until the mid-1400s. Prior to the arrival of the Māori, moa filled the vacant large herbivore niche, as mammals and marsupials failed to reach New Zealand. The moa dominated New Zealand's ecosystems for tens of thousands of years, their only predator being the Haast's eagle which also went extinct in the 1400s as their prey, the moas, disappeared.

Box

Easter Island

In perhaps the most famous example of indigenous human-triggered mass extinctions after just 450 years of the Rapa Nui civilization, seabirds and indigenous land birds were expiated from Easter Island by 1450 CE and were not found in excavated Rapa Nui garbage dumps thereafter. The extermination of these bird species occurred because pigs or goats did not make the long sea voyage that led to the colonization of Easter Island. Instead, the Rapa Nui obtained protein from seabirds, land birds, and porpoises, as well as from chickens which they had introduced. This, coupled with the lack of trading partners, as Easter Island is the most isolated island in the Pacific, meant the Islanders were on their own. Instead, the Rapa Nui concentrated on transforming the island through deforestation to gardens for cultivating potatoes, yams, taro, bananas, and sugarcane. They also transformed the landscape by building monumental platforms and statues, utilizing palm logs to transport the increasingly large moai from the highlands to the coast. As a result of both enterprises, Easter Island was completely deforested by 1450 CE!

The deforestation also reduced the ability of the Islanders to utilize the limited marine resources. Easter Island lacks fringing coral reefs or a lagoon, which meant that fish and shellfish contributed little to the diet of the Rapa Nui. When the first Europeans ventured onto Easter Island, the few remaining canoes were small and leaked, and they were incapable of sustained ocean travel. The pace of deforestation also far outstripped the island's natural ability to reforest because of three factors. First, Easter Island is cool and dry, nearly temperate in nature. Second, Easter Island lies far from the fertile Pacific volcanic dust cloud. As a result, Easter Island's soils are poor and prone to erosion. And third, seed recruitment failed across the island as the abundant presence of rats ensured that most palm seeds (recovered from the island during this time of environmental stress at the peak of moai production) show evidence of rat tooth marks. By 1722, when Easter Island was first visited by Europeans, the Rapa Nui, the palm forest, and all the indigenous land birds were extinct, and most of the 30 or more breeding seabirds were no longer breeding on the main island.

Lastly, Jared Diamond suggests that Easter Island foretells the future of the earth; we have no intergalactic trading partners, and we are knowingly and willingly engaged in local, regional, and global environmental destruction because we are economically, politically, and culturally too inflexible to change our consumption habits.

What we can conclude from these examples is that pre-industrial humans had the capacity to implement extinctions, first, on the continents, starting with Africa where we evolved, and then radiating out among the other continents as humans arrived there. These extinctions followed early pre-industrial humans, as they first colonized islands adjacent to continents (e.g., Caribbean Islands, Balearic Islands, Madagascar, and the Channel Islands off the western US coast), and then much later as they developed technologies, for example, multi-hulled canoes, wave charts, and star navigation compasses, to sail to oceanic islands such as to New Zealand and Easter Island. According to Shepard Krech, the myth of "the ecological Indian," the ecological pre-industrial human, where indigenous peoples lived in perfect harmony with nature is just that—a myth.

3.4 Oceanic Island Extinctions—Post-1600

The invention of the marine chronometer by John Harrison in 1730 and subsequent improvements in time-keeping led John Arnold to mass-produce chronometers by 1783. Allied with the development of more accurate sextants and the new fetish to map everything, the golden age of maritime exploration was seeded and brought about the first Kondratiev wave which peaked in 1600. In fact, Captain James Cook found Hawaii by calculating longitude, using a modified Harrison clock in 1778. This period that began in the mid- to late 1400s was led by the British, Dutch, and Spanish and initiated the expansion of European colonialism around the world. Although colonial trade, European hunting for trophies (◘ Fig. 3.2), deforestation, and expansion of European agricultural mono-cropping across the Americas, Africa, and Asia threatened and endangered many animals, few were

3

Fig. 3.2 Tiger hunting in British India, 1903. The couple are George Curzon, 1st Marquess Curzon of Kedleston and Viceroy of India, and his wife, Mary

driven to extinction. For instance, pre-European Indian tiger numbers have been estimated at 40,000, yet ~3500 still survive in modern India today.

Instead, in this period of colonial expansion and the golden age of exploration, extinctions were mostly restricted to oceanic islands along ship routes to newly discovered and conquered colonial lands. For instance, Mauritius, Réunion, and Rodrigues in the Mascarenes, the Seychelles in the Indian Ocean between the Cape of Good Hope in southern Africa and India and Indonesia, Saint Helena, Tristan de Cunha and Ascension Island in the south Atlantic, and Bermuda all lost significant numbers of birds and animals. These extinctions were driven by ecosystem destruction, competition for resources by introduced exotic species, mostly pigs and goats, predation by introduced species, and introduced diseases. In the case of ecological destruction, it is not just the destruction of ecosystems (e.g., the clear-cutting of forest), but the collapse of ecological services, the fragmentation of ecosystems that reduced gene flow, and trophic cascades whereby insufficient resources are left isolated in patches that could not support tertiary consumers. Above all else, the hunting of animals for food by humans, particularly sailors and later island colonists, contributed to these extinctions.

Sadly, these island extinctions are common. Oceanic island animals are typically rare, geographically highly localized, specialized, and hence more prone to extinction. According to both Soulé and Quammen, there appear to be 18 factors that contribute to extinction, and most, if not all, of these factors contribute together, or in isolation, to island extinctions. These are in no particular order: rarity and low density, rarity combined with animals being patchily distributed, limited dispersal ability, inbreeding and loss of gene heterozygosity, small gene pools associated with the founding population, hybridization among closely related species, all manner of habitat loss including long-term environmental trends, successional loss of habitat, habitat disturbance and habitat destruction, some kind of catastrophic event such as a hurricane or drought, extinction or reduction of mutualist populations such as Hawaiian honeycreepers and the plants they help pollinate, increased competition and predation, introduction of a new disease, and finally hunting and collecting.

Evolving in isolation, free from predators and continental diseases, puts these animals at a distinct disadvantage once humans arrive. Humans bring diseases and exotic animals, both herbivores (e.g., pigs and goats on most Indian and Pacific islands, deer in New Zealand) and carnivores (e.g., stoats in New Zealand and pigs and snakes in Guam); in addition, they rapidly reduce resources through overhunting, overfishing, cutting of forest to open up forest for cultivation, and removal of top soil through overgrazing, over-compaction, or sod-removal for fuel. The islands of Hawaii, Mauritius, and New Zealand illustrate a number of these 18 extinction factors at work.

▶ **Example**

Hawaii

In **Hawaii,** feather collecting (and by its very nature, the killing of the birds) by indigenous pushed many bird species to the edge of extinction, even exterminating several species, while the recent introduction of avian malaria and avian pox have triggered more recent extinctions. According to LaPointe, the introduction of rats also contributed to Hawaii's extinction event along with the loss of forest habitat, in particular, by the expansion of agriculture in the form of sugar, pineapple, and coffee plantations; all of these contributed to extinctions before the 1800s. Since 1800, 18 bird species have gone extinct, and today, few endemic bird species occur below 5000 feet where avian pox and malaria is unchecked. Feral pigs also pressured endemic birds, both by increasing the habitat for mosquitos by eating the pith of tree ferns and leaving hollows to fill up with water and mosquito larvae and by eating eggs of ground-nesting birds. All told, greater than 50% of the 57 known Hawaiian honeycreepers have gone extinct. Avian malaria was probably introduced with the introduction of exotic and cage birds in the early 1800s, while avian pox might have arrived with migratory birds. During their isolation and evolution, the honeycreepers were particularly vulnerable to these common continental diseases, having lost their natural immunity. ◀

▶ **Example**

Mauritius

The dodo (*Raphus cucullatus*) went extinct in a very short time because of the introduced pigs, left as a future food resource for Dutch mariners in the late 1500s and early 1600, and hungry Dutch sailors. The dodo and close relative, the extinct Rodrigues solitaire (*Pezophaps solitaria*) split from a common ancestor about 25 million years ago, but when both evolved flightlessness is not known. Their closest flying

3

relative is the rather large, island-specialist, the Nicobar pigeon (*Caloenas nicobarica*). This pigeon nests on small offshore islands from India to Australia, but frequently feeds on the mainland. This facet of its behavior probably led it to colonize Mauritius accidently; but once there, the opportunity to colonize other islands, other than the Mauritius group, was very limited. Isolated on these two islands, the ancestors of the dodo and solitaire became subject to environmental pressures that selected for bigger and bigger birds (ultimately sacrificing the ability to fly) so they could get through periodic droughts and famines. This solution to climatic variability on tropical islands was their undoing as Dutch sailors and their pigs ate either them or their eggs. ◄

Box

New Zealand

The arrival of the Māori ended the reign of the moa in New Zealand through hunting, but the Māori also pushed 16 other bird species into extinction. These include the adzebill, a large flightless bird not related to moa, two raptors, a swan (but also found on the Chatham Islands), a goose and three species of duck, a raven, an owlet-nightjar, and a penguin, among others. The Europeans and their deliberate and accidental introduction of alien, exotic animal and plant species doomed much of the rest of the native flora and fauna of New Zealand to extinction. In fact, 14 bird species scattered in the North, South, and Auckland Islands were exterminated directly or indirectly by European settlers.

Today, some 76 endemic land bird species survive, many of them on some 200 mammal-free islands adjacent to either the North or South Islands, while another 17 seabirds are endemic breeders. Of the 76 endemic land bird species, four are critically endangered, including the Kākāpō, the world's largest flightless parrot, the Malherbe's parakeet, the Okarito brown kiwi, and the South Island kōkako, while another 14 are endangered. New Zealand's list of introduced exotic pest species is long and their numbers huge. It is estimated that there are 30 million brushtail possums on the main islands, many millions more rats, but also large numbers of stoats and domestic and feral cats. According to Brian Owens, mammalian pests cost the New Zealand government NZ$70 million each year, while NZ$3.3 billion are lost in lost productivity each year. In response to these costs, the New Zealand government announced in 2016 that they would eliminate all vertebrate predators (rats, brushtail possums, and stoats, among others) by 2050 from New Zealand to protect is critically endangered endemic species at a cost of NZ$6 billion. Although audacious, more than 1000 islands across the world have been cleared of invasive species, among these include 200 New Zealand islands. Inhabited islands have proved more difficult to free from invasive animals; nevertheless, in 2011, after a four-year program, Rangitoto and Motutapu islands off the coast of Auckland were declared invasive predator-free; although visited by tourists, these islands have remained invasive predator-free.

Together, the once free-ranging red and fallow deer, now mostly constrained and bred for their meat, and millions of sheep continue to graze in unnatural park-like landscape of grasslands across much of New Zealand where once there were millions of hectares of native forest.

▶ Example

Chatham Islands

It is thought that on the 966 km^2 of the Chatham Islands, lying some 500 miles east of the South Island of New Zealand, some 22 bird species have gone extinct: they include a penguin, a swan, a shelduck, several *Pterodroma* petrels, several flightless rails, a fernbird, a bellbird, and a crow. Polynesians started visiting the Chathams regularly some 400–450 years ago, taking birds and possibly seals for food. After 1791, Europeans hunted whales and seals in and amongst the islands, but farming and permanent human settlement that began in 1892 sealed the fate of most of the now extinct birds. The introduction of sheep, goats, and rabbits, and then cats to control the rabbits, eradicated the native scrub vegetation. The cats also went after the birds, as did commercial bird collectors. The hunting, predation by alien species including rats, mice, pigs, possums, hedgehogs, cats, and dogs, and deforestation pressures on such a small island group were simply too much, in too short a time, for the birds to cope up with, making the Chathams' one of the worst examples of human-caused extinction events in history. Furthermore, the introduction of cattle, pigs, goats, sheep, and horses has reduced the indigenous vegetation to small scattered fragments. ◀

▶ Example

Guam

Sadly, Guam is the most recent example of an island extinction event via the accidental introduction of the brown tree snake (*Boiga irregularis*). The brown Tree snake is a poisonous, arboreal snake native to the western Pacific from Indonesia to Australia. First observed, and possibly introduced amongst cargo from Papua New Guinea, in the 1950s on Guam, the snake has gone on to exhibit densities in excess of 5000 snakes/km^2. The snake has been directly responsible for the extinction of 13 native Guam forest birds, and only 2 species, the Guam rail or ko'ko' and Guam kingfisher, survive in captive breeding programs that have released some birds back onto snake-, cat-, and dog-free islands. The subsequent future effects are going to be critical, as 60–70% of the island's native trees were dispersed by these birds. ◀

Box

Australia

Not an island as defined by humans, rather a continent, but in reality, it is a big island with a high degree of endemism due to Australia's geologic voyage across the southern oceans after detaching from Antarctica 85 million years ago. The endemism and maintenance of a marsupial mammal fauna placed it at a significant disadvantage when humans, dingoes, and other human-assisted placental mammals (e.g., red foxes, feral cats, European rabbits, goats, pigs, donkeys, dromedary camels, horses, and water buffalo) arrived, starting about 65,000 years BP. Without a placenta, a fetal marsupial is born smaller and less mature than placental mammals. It is

3

unable to grow as fast, because the nutrients it receives from the mother are limited through its more primitive umbilical cord. However, some marsupials such as the opossums compete with placental mammals without a problem as evidenced by their recent invasion of North America. Nevertheless, according to John Woinarski and coauthors, 100 endemic Australian species (including 34 marsupials and nine birds) living in 1788 are now believed to be extinct. Woinarski and coauthors suggest that the marsupial extinctions were due to the introduction of predators and the poisonous cane toad; plants by deforestation; frogs by disease; reptiles by the introduction of a snake; and invertebrates by a combination of human-induced changes.

Prior to 1788, native Australian fire management practices, called fire-stick farming, were spatially limited, created a mosaic of habitats, and were sustainable. Sadly, since 1788 and the colonization of Australia by the British, extensive deforestation around the coastal fringe of Australia to provide agricultural land and urban and suburban sprawl has changed the way Australia's native bushland responds to fire. The land has dried, warmed, and became more prone to fire (see Threshold ▶ Chap. 4 for explanation). Among the few native successes, several parrot species such as the galah and the rainbow parakeet, the deadly brown snake, and the funnel-web spider have colonized the suburbs. But many marsupials, birds, and plants have become isolated in native forest remnants. This isolation has proved their undoing during the 2019–2020 fire season. Unable to escape their isolated native forest and bushland remnants, hundreds of millions of animals, birds, reptiles, insects, and plants have been killed, and this on the back of a six- to seven-year long drought in the interior of Australia. Upward of 25,000 koalas have been killed—they are for all intent-and-purposes *functionally extinct*! The highly endangered and nomadic regent honeyeater down to no more than 400 individuals might be wiped out. According to Graham Readfearn and Pat Hodges, the Kangaroo Island dunnart, a tiny mouse-like marsupial found nowhere else, is likely gone as its island was burnt.

The 2019–2020 fire season was exceptional but not unexpected! Scientists have warned people and politicians for decades that global warming coupled with deforestation and mismanagement of Australia's bushland primed the continent for this catastrophe. As of January 7, 2020, some 23,000 square miles of land was burnt—an area the size of the US state of West Virginia! This is nearly twice the area that burnt in the Brazilian Amazon in late 2019. Ancient old Gondwana forests and wetlands in Queensland and NSW are burning as the conflation has contributed to 40 °C days. What's more, the World Wildlife Fund-Australia estimates that by mid-January 2020 a total of 1.25 billion animals died due to the fires are based on Chris Dickman's 2007 report for WWF that estimated animal population densities across New South Wales.

But this is not half the story—once the fires are extinguished, what will the animals, birds, reptiles, and insects do for food. The fires are burning much of the remaining native forests and bushlands. The remaining small areas of native forest and bushlands will support fewer species, and fewer individuals of each species as smaller areas have less habitats, fewer niches, and less habitat and environmental variety. By clear-cutting and fragmenting the forest and bushland landscape for agriculture, introducing exotic mammalian and reptile predators, mining and burning coal for electrical power generation, and sprawling into the agricultural and native landscape, Australians have contributed to this crisis locally and globally. For instance, Australian mining companies are among the biggest global contributors to GHG emissions.

3.5 Coastal Islands near Continents

Easily accessible coastal islands such as Tori-shima and Minami-Kojima Islands near Japan; Tasmania off Australia; and Papa Westray and St. Kilda off Scotland; Grimsey, Eldey, and Geirfuglasker Islands off Iceland, Funk Island off Newfoundland, and Bird Rocks (Rochers-aux-Oiseaux) in the Gulf of St. Lawrence have, over the centuries, offered rich hunting grounds for feather hunters, "vermin eradicators," egg hunters, and meat hunters. And adjacent to South America in the Humboldt current, islands rich in seabirds have offered plenty of ammonia for guano-gathers.

Box

Feather hunters killed in excess of five million short-tailed albatrosses between 1885 and 1903 on Tori-shima Island, and this harvest continued into the 1930s when 3000 were killed between December 1932 and January 1933 (U.S. Fish and Wildlife Service 2008). Such was the killing that although hunting was banned in the late 1930s by Japan, the albatross was thought to be extinct by 1949. However, 25 birds and six pairs were observed on Tori-shima Island in 1954. Today, at least 4200 short-tailed albatrosses of various ages are scattered across the northern Pacific. Although volcanic activity continues to threaten the principal breeding colony on Tori-shima Island, the world population of short-tailed albatrosses is growing at 6.5–8% per year. This is remarkable given the continued threats from long-line fishing by-catch in the northern Pacific.

Similar to the extraordinary efforts marshaled across Eurasia and North America to eradicate the gray wolf because it was designated as vermin, the Tasmanian wolf or thylacine was also classified as vermin, but sadly in this case was completely exterminated. The last known thylacine was kept in Hobart Zoo and died in 1936. Whereas the gray wolf continues to survive in remote wilderness and was recently reintroduced into Yellowstone as its critical role in ecosystem services has been documented, the scarce thylacine confined to Tasmania was

3

unable to escape and was exterminated before modern ecological analyses could support conservation efforts.

The northern Atlantic's penguin equivalent, the great auk—big, flightless, and vulnerable—was hunted to extinction across three countries for its feathers, eggs, and meat. The last great auks on Papa Westray and St. Kilda off Scotland were killed in 1840, and the last ones on the islands of Grimsey, Eldey, and Geirfuglasker off Iceland were killed in 1844.

Indonesia is home to rapidly dwindling tiger and rhino populations. While in 1970, 1000 tigers survived on Indonesia's island of Sumatra, today there are less than 400 left. Sadly, Javan Tigers disappeared in the mid-1970s, while Bali Tigers survived through until the 1950s. Javan rhinos are the most critically endangered of the five rhino species, with only 63 individuals left in Ujung Kulon National Park on the island of Java. Both tigers and rhinos were, or are being, driven to extinction by deforestation, and hence a loss of habitat and poaching. The last Javan rhino in Vietnam was poached in 2010. Elsewhere, the spread of palm oil plantations across Borneo (see Agriculture) has led to the loss of 150,000 orangutans.

3.6 Continental Extinctions

Over the last 150 years, "the island extinction disease" has spread to the continents and their adjacent islands, and this is critically related to the species-area relationship. This relationship simply states that small areas harbor fewer species, and fewer individuals of each species. Less area also means less habitat, fewer niches, and less habitat and environmental variety.

When comparing the dynamics of extinctions among islands and continents, it is perhaps easier to visualize using the concepts of isolates and samples identified by Preston and reexamined by Quammen in his book *Song of the Dodo*. A sample is a group of species or a parcel of landscape that exists as part of a greater whole; in contrast, an isolate is a sequestered group of species or an isolated parcel of landscape. In isolates, species are rarely, if ever, reinforced by other members of the same species wandering in from the outside. It is thought that approximately 2–3 million years ago, the first Galapagos finch arrived, survived, and bred, seeding a line of 14 new Galapagos finches. Since this one stochastic occurrence, no new finches have followed. In this case, as in other oceanic islands, the few lonely individuals were utterly on their own. On the mainland, a small area is a sample of a vastly larger area, with the net result that more species exhibit lower thresholds for extinction. This is MacArthur and Wilson's equilibrium theory of island biogeography. Even locally, rare continental species are maintained by individuals wandering in from the larger landscape. But if the sample is truly isolated such as in sky islands (e.g., those mountains in the Great Basin of Nevada surrounded by deserts), wandering individuals no longer arrive, and then extinction is a very real possibility. This is Brown's non-equilibrium theory of island biogeography. This is especially true of continental species that have restricted geographic ranges, in other words, are endemic to particular continental areas, are dietary or habitat specialists, have a large body size over 100 kg, are prone to human hunting and collecting activities (e.g., rhinos for horn and elephant for ivory), and are tertiary consumers (otherwise known as an apex predator such as lions or cheetahs), with low population density at the top of the food chain.

So, as we create more isolates on the continents through hunting and collecting and ecological destruction and habitat fragmentation, fewer species can survive, and those that do survive struggle as fewer wandering individuals support their losses and contribute fewer genes to these isolated samples. A powerful and very sad example of this is from North America. Since 1970, North America has lost 3 billion birds or a 29% reduction from 10 billion to 7 billion birds. Reporting in the journal of *Science*, lead author Ken Rosenberg suggests this signals a widespread ecological crisis, and one that will get worse unless North Americans reduce sprawl, reduce the use of neonicotinoid pesticides and the loss of insects, stop free-roaming domestic cats, and work to abate global climatic change. Most worryingly, declines in common species including sparrow, swallows, warblers, and finches across all habitats were observed, even those that have embraced the built environment. Nineteen common species have lost more than 50 million birds each, mirroring losses in Europe of both starling and house sparrows, species once thriving in the built environment. These losses also mirror massive losses observed among insects and amphibians.

Species losses (or extinctions, be they local or global) are both structural and functional. Working in the Andes, Isabel Donoso and colleagues found that a 10% reduction (or downsizing) in the total number of avian frugivores (seed/fruit eaters) led to a 40% reduction in long-distance seed dispersal, yet only a 10% reduction in the number of frugivore species. A 40% reduction in seed-dispersal will, just like Barro Colorado in Panama, lead to forest and ecosystem decline and ultimately ecosystem death. The ecosystem services that frugivores provide simply cannot be overstated.

In the tropics, road-building seeds deforestation through a vicious feedback cycle. As mentioned in ▸ Chap. 5 under the title "**Slash-and-burn and income inequality,**" road construction at the forest frontier is used by many governments to relieve urban population stress. As a result of road-building, deforestation and low-density agricultural sprawl within tropical forest environments result in a plethora of land-use cascades. This is because in subsistence agricultural system, survival is critically important, yet with no cash and little or no means to invest in fertilizer, families remain trapped in abject poverty. However, optimism reigns, so long as wildlands (e.g., uncut forest) remain, these are seen as opportunities by people who have no other alternatives. As a result, millions of slash/burn subsistence farms cut forest, cultivate yams or maize or bananas until the soil fertility gives out, and then wealthy cattle ranchers frequently step-in offering money for the degraded land to the destitute farmers, who then move down the road to the new forest frontier and cut a new section of forest. This is repeated until the forest disappears.

Sadly, the poor quality of tropical forest soils due to the combined effects of high temperatures and lots of precipitation that strips the soil of nutrients combines with the income inequalities to drive this land-use cascade. Nearly all of the tropical forest nutrients are found in the vegetation. The root systems of tropical forests target the top 20–50 mm of soil where the nutrients survive in the root litter; the rapid uptake of these nutrients in this near-surface soil environment means that once the forest is cut and the slash burned, little or no nutrient capacity is left in the soil. Adding artificial fertilizers speeds up the soil nutrient loss after forest removal such that each subsequent year requires more and more fertilizer per square meter of soil until the cash and food returns fall below the costs of fertilizer inputs. Once this threshold is crossed, this triggers land abandonment. At this point, wealthy cattle ranchers step in, purchase the land cheaply, and seed the degraded land with Africa Zebra grass allowing some nutrients to be extracted for their cattle. But in this environment, cattle densities are limited to one cow per ten to twelve hectares, making this a non-profitable option for small-scale subsistence farmers. The result is a mosaic of tropical forest patches surrounded by an ever-expanding rural low-intensity agricultural sprawl that isolates forest species, condemning them to local extinctions. Paths and roads do more than isolate and bisect forests and their ecological communities. Many forest species fear open spaces because light gaps are preferred habitats of both daytime and nighttime predators and provide access to dense forest for brood parasites such as cowbirds, cuckoos, and indigo birds in Africa. Brood parasites are bird species that lay their eggs in the nest of other bird species. Bird densities are reduced up to 1000 meters from a forest bisecting road. Recall as well that at a global scale, tropical deforestation aided and abetted by road-building and low-density agricultural sprawl accounts for 20–25% of all GHG emissions. Most of these emissions occur during deforestation, but substantial emissions occur when agriculture lots are converted and sustained as cattle pastures.

Today's pace and process of extinction is frequently referred to as both competitive and indirect *attritional overkill* rather than *associated blitzkrieg overkill* that describes the Pleistocene megafauna extinctions.

> *Attritional overkill* requires human presence in the area, human manipulation of their environment, and alteration of ecologies through farming practices, and hunting and collecting. The two types of *attritional overkill* are where humans have an indirect impact—this is "man-the-farmer and habitat manipulator" (i.e., where human activities stress local fauna)—and are where humans are the competitor—humans competitively exclude other animals due to an overlap in diet preference, feeding strategies, or habitat utilization.

This last set is more applicable to a modern last 5000-year period since the advent of agriculture and human permanent settlement beginning in present-day Iraq. Since the advent of agriculture and associated deforestation, even abundant, widespread continental species have gone extinct. The most recent wave of extinctions moved to the continents and their adjacent islands in the mid-1800s, as illustrated by North America.

American Extinctions

These extinctions include what most ornithologists consider the most numerous birds ever: the passenger pigeon (*Ectopistes migratorius*). Audubon in 1813 observed 160 flocks in one day and estimated 1.11 billion birds passed overhead. But by 1914, the last passenger pigeon died in captivity—a combination of overhunting, deforestation, and an introduced disease, the chestnut blight, that killed most of the American chestnut trees, which then represented 25% of all the trees in passenger pigeon's home range from Maine to Georgia. Sadly, the passenger pigeon was a dietary specialist, consuming American chestnut tree mast-nut. Its huge flocks and colonial nesting strategies were an evolutionary adaptation to find and exploit these resources. It is thought that once the birds declined in flock size to less than 500,000 members, they were unable to locate sufficient nuts that were scattered, but spatially clustered, across the American East.

Other North American losses include the Carolina parakeet, the ivory-billed woodpecker that might have survived in Cuba until the early 1990s, the Bachman's warbler, a bamboo specialist, the great auk, and the Labrador duck. In the case of the God-bird or ivory-billed woodpecker, it was driven to extinction by the cutting of southern old-growth cypress and the introduction of European honeybees, which possibly out-competed the ivory-billed woodpecker for nest-holes. Recent evaluation of the Carolina parakeet's genome confirms that hunting and deforestation by European humans were responsible for its rapid demise. According to Pere Gelabert and coauthors, inbreeding prior to European settlement could not be identified, suggesting neither native American hunting pressure nor environmental changes contributed to its demise.

These broad scale species-ending extinctions are also accompanied by smaller-scale local extinctions. For instance, for the damming of the Chagres River in 1914 to form Gatum Lake as part of the Panama Canal, a remnant of tropical forest was isolated and cut off from the mainland to become Barro Colorado. Since 1910, the Smithsonian Tropical Research Institute has extensively monitored the island. Among the most surprising findings is that after the remaining jaguars swam away, seed predators (principally agoutis) were left unchecked, and as a result, Podocarpus trees were unable to seed and replace dead and dying trees. This is an example of a slow, inextricable ecosystem death.

3.7 Global Extinctions

With the advent of our oil-based civilization in the late 1890s, humans and the earth were set on a course to a sixth mass extinction. Recent work by Daniel Rothman examining the carbon cycle over the last 540 million years suggests that exceeding two critical thresholds in the carbon cycle trigger mass extinctions. One, extinctions occur over geologic time scales if changes to the carbon cycle occur faster than terrestrial and marine ecosystems can adapt. Second, extinctions occur over shorter timescales, over hundreds to tens of thousands of years, if there is a sudden, huge amount of carbon added to oceans, something we are witnessing today. He calculated that 310 gigatons of carbon are needed to be added to exceed this second threshold—this is the amount he estimates we will add to the world's oceans by 2100. At this point, a sixth mass extinction is inevitable, which will play out over the next 10,000 years or so.

This scenario is already playing out today. Polar bears are starving to death across the Arctic Ocean and Arctic sea ice as sea-ice margins are in serious retreat. Satellite imagery, ice core data, permafrost records, and tree-ring evidence all suggest that Arctic sea ice coverage is declining at its fastest rate in the last 1500 years. In 2017, the maximum Arctic sea ice extent was the lowest yet recorded by NOAA's Arctic Report Card.

The first officially recognized mammal extinction caused by recent human-induced climate change has been identified—a small rodent, the Bramble Cay melomys. It formerly occupied a tiny island in the Torres Strait located between Papua New Guinea and Australia. First observed in 1845, the melomys experienced a catastrophic decline from several hundred individuals observed in 1978, to the last seen in 2009. A trapping survey located none in 2014. The culprit: sea-level rise! According to Brian Howard, since 1998, rising sea levels shrunk the melomys' vegetation habitat above high tide by 97%, leading to the melomy's extinction.

3.8 Emerging Crisis of Newly Threatened Species

The saiga antelope has undergone several population boom-and-bust cycles since the mid-1800s as demand by Chinese medicine for its horns has fluctuated. In the last 15 years, 95% of the saiga antelope have been exterminated by hunting and disease. The hunting has focused on their valuable horn; today, saiga horn goes for $4600/horn which is more than the value of either rhino horn or helmeted hornbill castes. In 2010 and again in 2015, 12,000 and 120,000 saiga, respectively, were found dead, apparently from *pasteurellosis* and/or *clostridia*, which are infectious diseases of the lungs and intestines (Nichols 2015). In 2015, the 120,000 dead saiga represented 50% of the world's population. Only some 50,000 survive in Kazakhstan and in two small isolated areas in Mongolia.

The combination of significant and sustained hunting and two random stochastic disease events suffered by the saiga antelopes illustrates the synergistic nature of many near extinctions or extinction events. Habitat destruction or hunting, if severe enough, can precipitate population crashes, while the final straw that breaks the camel's back is typically some synergistic feedback process; Brook and coauthors suggest that saiga antelopes might have had a near extinction event after a disease nearly decimated their surviving herds.

Tropical deforestation creates a series of synergistic feedback processes that increasingly threaten tropical forest species. Deforestation reduces habitat quantity and quality, increases fragmentation, and increases human access points throughout a forest that facilitates more hunting. Hunting directly increases species mortality, while fragmentation increases the vulnerability of a species to invasive species, parasites, and predators. Hunting can also remove apex predators, initiating ecological cascades that increase the number and diversity of seed predators, reducing forest recruitment, and reducing habitat quantity and quality (e.g., BCI and the jaguar). According to Brook and coauthors, deforestation also leads to desiccation of the forest that reduces habitat quantity and quality, increases invasibility, and increases fire penetration that increases species mortality and further reduces habitat quantity and quality.

In southern California, San Joaquin kit fox exemplifies this process, as it is struggling due to the synergistic combination of habitat destruction and fragmentation associated with the expansion of agriculture and climate change as the San Joaquin valley has become drier.

Box

Elephants and Rhinos

Elephants and rhinos have suffered human encroachment on their lands and have been poached for their horns and tusks for millennia, such that today many rhino species are critically endangered. Only 72 Javan rhinos and less than 80 Sumatran rhinos remain. And according to Save the Rhino, black rhinos declined from more than 70,000 individuals in 1970 to less than 2410 in 1995, a 96% die-off in 20 years. In contrast, southern white rhinos have rebounded from their near extinction in the early twentieth century to approximately 18,000 individuals, with the majority living in South Africa. And the Indian or greater one-horned rhino has also seen miraculous recovery from less than 200 to over 3580 individuals in India and Nepal due to extensive and coordinated conservation efforts. Nevertheless, today illegal poaching of southern white rhinos is unsustainable, and rhinos are disappearing rapidly! Even South Africa, a relatively wealthy society, with a strong tradition of wildlife conservation and a large number of well-protected reserves, is struggling to maintain its rhino populations. Between 2008 and 2018, Mateo-Tomas and López-Bao report that 7899 of South Africa's 20,306 rhinos were poached! Elsewhere, in 2008–2015, 210 rhinos were poached in Kenya, 130 in Namibia, and 436 were poached in Zimbabwe!

African elephant species are also threatened by poaching; while the Indian elephant numbers appear stable, the Sri Lankan subspecies is struggling due to land-use conflicts, and elephant numbers in Sumatra and Borneo are quite small and shrinking. In sub-Saharan Africa, the recent postcolonial (mid-1960s) economic instabilities and armed conflicts have encouraged a recent influx of modern assault rifles into the region, and together with increased rural-urban inequalities, they have facilitated an acceleration in illegal poaching of both rhinos and elephants. Prior to the early 1990s, traditional markets for rhino horn (the Horn of Africa) and elephant tusk (China, Thailand, Vietnam, and Cambodia) kept up a steady rate of illegal poaching; the recent post-1991 surge in globalization has opened more routes for poachers to get their illegal poached

3

items to their markets in the Horn of Africa and East and Southeast Asia, fueling a surge in poaching. For instance, in 1976, Douglas-Hamilton estimated there were 1.34 million elephants inhabiting some 7.3 million km^2. Today, elephant numbers have crashed to around 415,000 animals in over 1.9 million km^2! Wittemyer and coauthors estimate that poaching kills upward of 30,000 elephants a year. At this rate, wild elephants will be extinct in Africa within 15 years!

However, at least in South Africa's Kruger National Park, poaching has declined since 2018, with the introduction of free-running packs of American coonhounds. Working with the Ivan Carter Wildlife Conservation Alliance and dog-breeder Joe Braman, Kruger NP has captured 145 poachers or approximately 54% of all known Kruger poachers and confiscated 53 guns (Engkilterra 2019). Kruger's standard leashed K-9 anti-poaching units prior to the arrival of the coonhounds typically caught only 3–5% of all Kruger poachers. The unleashed dogs work as a team with men on the ground and helicopter pilots in the air. Together, they protect each other, but they are more mobile, and when the packs confront poachers, the poachers typically abandon their weapons and climb a tree. Subsequent arrests are relatively peaceful and safe affairs.

Sadly, giraffes are also disappearing due to poaching, as they are sought for their meat, bones, hides, and tails to use as whisks. Mateo-Tomas and López-Bao suggest that over the last 30 years, giraffe numbers have decreased by at least 36%.

► Example

Sharks, Rays, and Chimaeras

Dulvy and coauthors estimate that 25% of all sharks, rays, and chimaeras are threatened, and of these, large-bodied, shallow-water species are at most risk of extinction, especially rays. According to Myers and coauthors, annual shark surveys since 1972 off the North Carolina coast suggest sandbar sharks have declined 87%, blacktips 93%, tiger sharks 97%, scalloped hammerheads 98%, and bull sharks 99%. However, it appears that the oceanic whitetip shark is most at risk. It has recently suffered a 95% die-off in the last 30 years and is nearing extinction. The reason for this rapid die-off is that oceanic whitetips are caught in large numbers as a bycatch in pelagic (deep-water) fisheries—albatrosses suffer a similar fate. Both are snagged by pelagic longlines, pelagic gillnets, handlines, and pelagic and bottom trawls. In addition, whitetips are actively pursued by the shark-finning industry that, according to Worm and coauthors, killed 97 million sharks in 2010. They estimate that average shark exploitation rates exceed 6.4%/year, and this exceeds average rebound rates among all sharks of 4.9%/year; therefore, shark-finning is totally unsustainable. Boycotts of restaurants selling shark fins are having an effect, but maybe too late. The trophic cascades that this mass die-off is triggering are only just being appreciated. For instance, the loss of large-bodied, shallow-water shark species has triggered a population explosion in smaller sharks and rays off the Japanese coast, with the result that both wild and cultivated shellfish stocks have plummeted. In other words, one mass die-off has triggered another mass die-off. ◄

► Example

Amphibians

An even bigger and broader extinction crisis than that faced by Indian vulture species is, according to Stuart, the current global amphibian extinction crisis first reported in the 1950s. Today, 43% of amphibian species face extinction, which is up from 32% in 2004, and yet no viable solution appears to be available. Culprits include habitat loss, especially tropical deforestation, climate change, and the global emergence of chytridiomycosis, an amphibian skin disease, following infection by either of two fungal pathogens: *Batrachochytrium dendrobatidis* (*Bd*) or *Batrachochytrium salamandrivorans* (*Bsal*). Chytridiomycosis has been blamed for the extinction or population decline of over 200 amphibian species worldwide. In fact, according to deer Sluijs, *Bd* first emerged in Australia in 1999, while *Bsal* was first described in the Netherlands among salamanders in 2013, following a 96% decline in salamander populations between 1997 and 2012. These two newly emergent diseases exhibit high transmissivity, high mortality, and no host immunity—these combinations are proving lethal to amphibians worldwide. ◄

► Example

Bees

Today, according to Whitehorn, neonicotinoids, including thiamethoxam in the UK, account for more than 25% of the global market share of insecticides, yet declines in bee populations, that are critical for agricultural production through their pollination services, have mirrored the expansion in use of neonicotinoids. Neonicotinoids appear to confuse bees and reduce their homing success—in other words, many bees struggle to return to their hives after exposure to neonicotinoids. A knock-on effect is that queen bees stop laying as the number of bees in a colony falls below a critical threshold. It would not be too far-fetched to state that we are living in the global bee extinction crisis. ◄

Although neonicotinoids have proved less deadly than organophosphates to birds and amphibians, new research suggests that declines in farmland birds throughout Europe are precipitating an ecological catastrophe. In France, bird numbers have declined 33% in just the last 15 years, and according to Hallmann and coauthors, flying insect numbers have dropped by 76% in Germany over the last 27 years across all types of terrain and ecosystems; overall, farmland birds have declined in excess of 50% in the last four decades. In addition, they also identified that flying insects suffered an 82% decline by mid-summer. This "insect apocalypse" is particularly troubling because a third of all human food crops are insect-pollinated, and over half a billion tons of insects are consumed by birds and mammals annually. If the losses found in Germany are replicated elsewhere, which it seems there are, the ecosystems services insects provide will be catastrophically and negatively impacted. And according to Amy Berenbaum, a human apocalypse will shortly follow an insect apocalypse unless we act now. She recommends the following:

» Get rid of your lawns... Lawns are total biological deserts. It's just grass with pesticide. The way it's grown does not support a lot of biodiversity. Leave leaf litter. The dead plant material is an important part of healthy soil and offers protection for insects during the winter,... and don't use bug zappers.

» Ruppenthal, A. October 2, 2019. Wttw PBS.

3.9 Attempts to Stop Further Extinctions

Attempts to stop further extinctions have typically been aimed at individual species, particularly those with name recognition and those animals that are cute or tug at people's hearts, for instance, the giant panda, whooping cranes, California condors, tigers, dolphins, and whales. In contrast, the pangolin (a scaly anteater) has not crossed into Western consciousness and so has failed to attract sufficient attention as global northerners have become immune to these critical life-and-death-extinction issues. The lack of public outcry in the Global North was very evident in the case of the death of the last northern white male rhino in March 2018, and in the plight of vultures in Africa and India.

Whenever attempts are made to save a critically endangered species or population, the future viability and vulnerability of either a species or population must be considered. Although a simple cost-benefit analysis of saving an animal is not necessarily ethically defensible, sadly it is still used. One approach to identifying the viability of a critically threatened species is to identify their minimum viable population where there is still:

1. a large, diverse gene pool,
2. a population sufficiently large that it can survive random stochastic events, for example, disease, flood, and hurricanes.

Soulé has suggested 50, where 50 is the minimum viable population having at least a 95% chance of surviving for another 100 years despite foreseeable effects of demographic, environmental, and genetic stochasticity and natural catastrophes. But this number has been found to vary for each species. For instance, the passenger pigeon's minimum viable population numbered maybe 10 million! On the other hand, the whooping crane declined to just 23 individuals, and the Californian condor was just 14 individuals.

Among islands, exterminating invasive alien intruders such as rats, cats, or pigs are the key to helping endangered animals, but alien eliminations also help all ecosystem services. In 2013–2018, 330 tons of rat poison was scattered over the island of South Georgia, near Antarctica. In May 2018, conservationists were able to declare South Georgia rat-free. The hope is that with the rats gone, the number of albatrosses, skuas, terns, petrels, and South Georgia pipits will rebound.

► Example

Whooping Crane

The combination of hunting and habitat losses meant that by 1941 just 23 whooping cranes were alive, 21 in the wild and two in captivity. In the early 1960s, the National Audubon Society (NAS), the Whooping Crane Conservation Association, Robert Allen—a NAS ornithologist, and Patuxent Wildlife Research Center began a captive breeding program. This was augmented by the International Crane Foundation and Calgary Zoo in 1976. By 2018, 163 remain in captivity, 431 breed in Wood Buffalo, Canada, and in winter in Aransas National Wildlife Refuge, while 163 reintroduced birds live wild in the eastern United States. Sadly, budget cuts by the Trump administration ended Patuxent Wildlife Research Center's 51 year program, though their flock of 75 birds were relocated to the Calgary Zoo and International Crane Foundation location in Wisconsin and to some private zoos where it is hoped private funding will secure the whooping cranes future. ◄

► Example

Californian Condor

In 1987, the last remaining free-flying Californian condors were removed from the wild, and for all intent and purposes, the Californian condor went extinct. However, this was not the end. Incredible efforts undertaken by the Peregrine Fund, the San Diego Wild Animal Park, and the Los Angeles Zoo meant that, by 1991, condors were being successfully returned to the wild. As of 2018, there were 488 condors, living wild or in zoos, all derived from 14 individuals. But this effort has required enormous financial and person-hour commitments, something many other species do not receive. Although saved from extinction, both these condors and cranes still suffer from habitat loss, low genetic diversity, power line collisions, and predation in the case of cranes, while poisoning of carcasses, illegal shooting, and nest site disturbance still kill condors. In both cases, very significant resources have been marshaled by committed environmentalists and will need to be committed for decades to come before we can feel secure about the survival of these two iconic species and conservation successes. ◄

► Example

Indian Vultures

Millions of vultures were first observed in the late 1990s dying from visceral gout, but it was not until a 2000 vulture survey that the decline became broadly evident. It is estimated that white-rumped vulture populations fell 99.7% between 1993 and 2003, while the Indian and slender-billed vultures declined by 97.4% (Prakash et al. 2012). At the peak of near-extinction, vulture die-off reached 40% per year. Parsees, adherent of Zoroastrianism, had also noticed the decline, as Parsees relied on sky burials by vultures in Towers of Silence to consume their dead. But it was not until 2003 that the painkiller and anti-inflammatory drug, diclofenac, given to cattle, was identified as the culprit by Dr. Lindsay Oaks (Oaks et al. 2004), working at Washington State University and sponsored by the Peregrine Fund. As diclofenac was subsequently banned by the Indian government in 2006, today, vultures are now being bred to return to the wild, as diclofenac has virtually disappeared from the wild. Created in 2001, the Jatayu Conservation Breeding Center (JCBC) at Pinjore began captive-breeding. Sponsored and assisted by the Forests Department and the Royal Society for the Protection of Birds, UK, the program has been a remarkable success, and in 2011, a consortium of NGOs, among them a Parsee group, and 14 Indian government agencies established seven new breeding centers based on JCBC's model. However, the low fecundity of these three vulture species means that they remain critically endangered and will not be seen in any numbers at the cattle dumps nor at the Towers of Silence for some time to come.

The rapid reaction of the Indian government once diclofenac was identified as the culprit is commendable, as it took less than three years to ban diclofenac, whereas in the USA, it took 20 years to ban DDT. However, the Indian government is not blameless—a law to protect sacred cows had encouraged the use of diclofenac to relieve pain among the sacred cattle, even those living on the streets. Later, another Indian law banned anybody from interfering with the once-threatened vultures, either alive or dead, meaning that researchers had to search for a disease culprit among Nepalese and Pakistani vulture populations. ◀

▶ Example

African Vultures

According to Ogada, Botha, and Shaw, ivory poachers are poisoning hundreds of African vultures a year. Poachers are poisoning elephants to both obtain ivory and contaminate elephant carcasses with poison to kill vultures so that there are fewer vultures circling over poached elephants and giving away their illegal activities. Between 2012 and 2014, Ogada, Botha, and Shaw identified 11 poaching operations across seven African countries that killed 155 elephants and 2044 vultures (killing vultures has a detrimental impact on ecological services; see Resources). They also observed that in two incidents vulture body parts were taken, something that is surprisingly common. In fact, Ogada and coauthors estimate that 29% of all vultures killed between 1972 and 2014 were associated with body part use in traditional medicines. In Nigeria, according to Saidu and Buij, 93% of traditional medicine dealers offered vulture bodies or body parts for sale in 1999–2011. As a result, vultures are in severe decline across Africa. During the last 30 years, according to Margalida and coauthors, eight vulture species, including white-backed, hooded, white-headed, lappet-faced, and Cape vultures have declined in numbers by at least 62%. In 20 months from January 2018 to September 2019, over 1000 vultures were poisoned, including 537 vultures killed in one instance in Botswana. ◀

Bird Decline in the USA Across the Global North, bird numbers have declined dramatically since 1970 as deforestation, agricultural intensification, sprawl, and the increased use of pesticides and herbicides have devastated bird populations. In the USA and Canada, Christmas Bird Count and eBird data show that three billion birds have disappeared, with the largest losses in the grassland habitats that used to exist in the Great Plains, boreal forest in Canada, and western forests of the Rocky Mountains. In the latter two areas, commercial timber extraction allied with climate change has conspired to greatly reduce bird populations. Twelve groups of birds have suffered the most, including American sparrows, wood warblers, blackbirds, Old World sparrows, larks, finches, swallows, nightjars, swifts, tyrant flycatchers, starlings, and thrushes. Old World sparrows have seen the greatest declines, as they have across Europe from whence they came in the late 1890s. In Europe, since 1980, house sparrows have declined by 62% and European starlings by 53%. According to Michael Gross, agricultural intensification cannot be completely blamed as EU edicts have reduced chemical (i.e., pesticide, herbicide) use in rural farmlands. But the use of the neonicotinoid family of pesticides since the 1980s, which persists in the environment, might explain *all* these losses.

At a broader scale, even though author Ken Rosenberg notes that since 1970, North America has lost three billion birds, some bird families have thrived due in large part to widespread habitat conservation and the banning of DDT. In fact, North America has gained 14 million woodpeckers, 15 million raptors, and 35 million ducks since 1970. Simply put, banning both DDT and the shooting of raptors have led to a revival in raptor numbers. In addition, Ducks Unlimited, North American Waterfowl Management Plan, and the creation of numerous National Wildlife Refuges have protected duck breeding and wintering sites throughout North America. The transition to forestry throughout the US South and maturation of trees planted within suburban housing developments of the 1950, 1960s, and 1970s across North America have greatly benefited woodpecker numbers. For instance, standing on Talmadge Bridge near Savannah, one is struck by just how much the urban/suburban/rural South is covered in trees (see ◘ Fig. 3.3). Just a hundred years ago, it was completely deforested.

◘ **Fig. 3.3** The Talmadge Memorial Bridge, spanning the Savannah River between downtown Savannah, Georgia, and Hutchinson Island (@Carol M. Highsmith 2017)

3

▶ Example

Albatrosses and Longline Fishing

In the last 30 years, Pardo, Sugden, and Tuck and their coauthors estimate that the numbers of wandering, black-browed, and grey-headed albatrosses have declined by 40–60% due to bycatch associated with longline fishing and climate change. However, there appears to be hope for albatrosses with the implementation of the Chilean system or trotline-with-nets developed to reduce depredation of toothfish by sperm whales and killer whales. The Chilean system increases the sink rate of the weighted line from 0.15 m/s to 0.8 m/s, making them inaccessible to albatrosses. According to Robertson and coauthors, the impact was immediate, for instance, Spanish vessels in 2002 killed 1555 black-browed albatrosses off the coast of Patagonia, but after converting to the Chilean system, Spanish vessels in 2006–2007 (and BirdLife International) reported no albatross bycatch. This reduction in mortality was also evident at breeding sites for both black-browed and grey-headed albatrosses on the Diego Ramirez and Ildefonso Archipelagos where black-browed numbers increased 52% and 18%, respectively, between 2002 and 2011. Although grey-headed numbers remained constant, at least no declines were noted. ◀

▶ Example

Fireflies

Fireflies are among the most adored and treasured insects due to their amazing and haunting light displays; however, their numbers are collapsing across the planet, and according to Sara Lewis, light pollution, habitat loss, and pesticides are contributing to this ongoing extinction event. Habitat loss, habitat fragmentation, and increasing human presence within critical habitat are the principal contributors to these losses. Urban/suburban sprawl increases light pollution that reduces the ability of fireflies to locate each other during the mating season, while habitat loss and fragmentation reduce firefly populations and isolate each other. Urban sprawl also increases pesticide loading, particularly in Global North countries where house owners and gardeners spray their lawns and gardens with much higher volumes of pesticides and herbicides. ◀

3.10 From the Brink of Extinction

New Species Incredibly given the voracious collecting by Victorians and subsequent collectors, scientists are still finding new species, while geneticists are evaluating the notion of what constitutes a species and are suggesting many subspecies should be elevated to full-species status. For instance, 71 new animal and plant species were discovered by California Academy of Science researchers. A research group from the Biodiversity Unit of Finland's University of Turka, in collaboration with Brazil's Instituto Nacional de Pesquisas da Amazonia, found 15 new parasitic wasp species in the Amazon rainforest and foothills of the Andes. In 2019, the World Wildlife Fund identified 163 new species from the Greater Mekong Region. In 2017, 28 new species of crustaceans were discovered in Antarctica. In 2019, the Natural History Museum announced it had identified more than 400 new species including 171 new beetle species. Researchers looking at cichlids in Lake Mweru identified 40 new cichlid species in 2019. Researchers also found ten new bird species and subspecies on islands just off the coast of Sulawesi, Indonesia, in 2019. Even in the USA, new species keep turning up; a new crayfish was identified by the Kentucky Department of Fish and Wildlife Resources early in 2020.

Rediscovered Species Global Wildlife Conservation recently launched a global project to relocate "lost species." In January 2019, the group relocated Wallace's giant bee (*Megachile pluto*) on the North Maluku islands of Indonesia. It had not been seen there since 1981. Wallace's giant bee is the world's largest bee, with a wingspan of 2.5 inches or 60 mm. The bee inhabits primary lowland forest, which is under considerable logging threat, and nests in tree-dwelling termite nests. The group also relocated several individuals of the Sehuencas water frog (*Telmatobius yurucare*), prior to this re-location—a single, very lonely male has lived in an aquarium in Bolivia. Sadly, 22% of all Bolivian amphibians are threatened by habitat loss, aquatic pollution, and climate change.

Incredibly, in early 2019, a "lost" Galapagos tortoise, the Fernandina giant tortoise (*Chelonoidis phantasticus*), was relocated on Fernandina. This species was last seen in 1906, although feces and evidence of feeding were located on the island's active volcanic slopes in 2015.

New Discovered Species Go Extinct The first individuals of the Rabb's fringe-limbed tree frog (*Ecnomiohyla rabborum*) were found in 2005 in Panama, but none could be relocated in a subsequent search in 2007. The last individual of this species named Toughie died in the Atlanta zoo in 2016.

In contrast to the news that a "lost" giant tortoise had been relocated on the Galapagos Islands, the first documented extinction of a bird species on the Galapagos Islands was announced in early 2019. In 2016, in what amounted to a first, Ore Carmi and coauthors identified a new species of vermilion flycatcher using molecular phylogenetic analysis, unique to the Galapagos Islands. But tragically, the flycatcher has not been seen in the wild since 1987.

Elevating Subspecies to Full Species Status Tim Crowe and Paulette Bloomer argue that identifying subspecies trivializes biologically different species. For instance, they argued in 2017 that the giraffe (*Giraffa camelopardalis*) is four morpho-genetic species rather than 11 subspecies identified using the biological species concept. This simplification is, however, a powerful conservation tool, as it clearly identifies that all four evolutionary significant units (ESUs) or species, the Masai, Northern, South African, and Reticulated, are in desperate need of conservation, rather than lumping them all together and suggesting that minor subspecies are threatened, but the species as a whole is not threatened.

Among birds, it is typically accepted that there exists between 9000 and 10,000 species. Barrowclough and coauthors suggest this is a vast underestimate, and using morpho-genetic evolutionary significant unit (ESU), they suggest there might be in fact 18,000 ESUs or bird species. They note that given "We have decided societally that the target for conservation is the species, so it follows then that we really need to be clear about what a species is, how many there are, and where they're found." As noted previously in this chapter, targeted conservation actions have and continue to save species from extinction. Hoffman and coauthors suggest that if in 1996 conservation efforts had ceased among the world's 235 ungulate species, today 148 species would be endangered and six species would have gone extinct. So, conservation is critically important at the scale of species. We suspect that rather than lamenting the loss of the last female and male northern white rhinos (*Ceratotherium simum cottoni*), more effort and funds would have been directed toward these beautiful creatures if they would have been classified as ESUs or full species.

Some Good News

Despite all the gloom of doom of recent extinctions, we do keep finding new species! Five new bird species and five new subspecies were recently found on the remote Wallacean Islands of Taliabum, Peleng, and Batudaka in Indonesia. In fact, Patrick Greenfield, in January 2020, reports that scientists have found 161 new species in the last 20 years. And we have on a rare occasion released back into the wild, once extinct species—the Guam rail being the most recent example. We are also seeing evolution among some bird species and the generation of at least new subspecies occurring at human timescales and not just among viruses and bacteria! Three bird species stand out: the European blackcap (see Urbanization), dark-eyed Juncos in the USA, and barn swallows in Argentina.

According to Vance, Price, and Ketterson, in the 1980s, a small group of dark-eyed juncos chose to stay on their wintering grounds around UC-San Diego campus. They were successful, nesting in trees and on buildings rather than on the ground, producing 3–4 clutches per season rather than 1–2. Males tended clutches more and became monogamous, male testosterone declined, intense male head and tail colorations dimmed over time, and their songs got higher in pitch, and they spread up to Los Angeles. Another similar group of juncos have recently settled on their wintering grounds in Ohio. Both examples are highly encouraging for their future survival as dark-eyed junco populations on their traditional breeding grounds have declined by 168 million individuals since 1970. In 1980, a number of barn swallows chose to stay on their wintering grounds around Buenos Aires. Winkler and coauthors found that in a very short time these birds had flipped their migration patterns, timing of their molt, and breeding seasons by six months. Today, this growing population of barn swallows are austral migrants, migrating to northern South America in their southern winter.

References

Benitez-Capistros, F., Hugé, J., & Koedam, N. (2014). Environmental impacts on the Galapagos Islands: Identification of interactions, perceptions and steps ahead. *Ecological Indicators, 38*, 113–123.

Brook, B. W., Sodhi, N. S., & Bradshaw, C. J. (2008). Synergies among extinction drivers under global change. *Trends in Ecology & Evolution, 23*(8), 453–460.

Barrowclough, G. F., Cracraft, J., Klicka, J., & Zink, R. M. (2016). How many kinds of birds are there and why does it matter? *PLoS One, 11*(11), e0166307.

Carmi, O., Witt, C. C., Jaramillo, A., & Dumbacher, J. P. (2016). Phylogeography of the vermilion flycatcher species complex: Multiple speciation events, shifts in migratory behavior, and an apparent extinction of a Galápagos-endemic bird species. *Molecular Phylogenetics and Evolution, 102*, 152–173.

Crowe, T., & Blomer, P. (2017). *A new approach to understanding subspecies can boost conservation. The conservation* (p. 2017). Published February 2. http://theconversation.com/a-new-approach-to-understanding-subspecies-can-boost-conservation-68364.

Dodson, C. H., & Gentry, A. H. (1991). Biological extinction in western Ecuador. *Annals of the Missouri Botanical Garden*, 273–295.

Donoso, I., Schleuning, M., García, D., & Fründ, J. (2017). Defaunation effects on plant recruitment depend on size matching and size trade-offs in seed-dispersal networks. *Proceedings of the Royal Society B: Biological Sciences, 284*(1855), 20162664.

Douglas-Hamilton, I. (1979). The African elephant survey and conservation programme. *Final Report to WWF/NYZS/IUCN,* Nairobi, Kenya.

Dulvy, N. K., Fowler, S. L., Musick, J. A., Cavanagh, R. D., Kyne, P. M., Harrison, L. R., Carlson, J. K., Davidson, L. N., Fordham, S. V., Francis, M. P., & Pollock, C. M. (2014). Extinction risk and conservation of the world's sharks and rays. *eLife, 3*, e00590.

Engkilterra, S. (2019). Texas pack hounds charge to the rescue of rhinos in South Africa, nabbing 145 poachers so far. *Nviro News, 18 Sept*, 2019. https://www.environews.tv/world-news/091819-texas-pack-hounds-charge-to-the-rescue-for-rhinos-in-south-africa-nabbing-145-poachers-so-far/.

Gelabert, P., Sandoval-Velasco, M., Serres, A., de Manuel, M., Renom, P., Margaryan, A., Stiller, J., de-Dios, T., Fang, Q., Feng, S., & Mañosa, S. (2019). Evolutionary history, genomic adaptation to toxic diet, and extinction of the Carolina parakeet. *Current Biology, 30*, 1–7. https://doi.org/10.1016/j.cub.2019.10.066.

Greenfield, P. (2020). Flycatchers and fantails: New songbirds discovered on tiny islands. The Guardian online Jan 9, 2020. https://www.theguardian.com/environment/2020/jan/09/flycatchers-and-fantails-new-songbird-species-discovered-on-tiny-islands-indonesia-aoe.

Gross, M. (2015). *Europe's bird populations in decline*. Elsevier.

Gusmão, A. C., Messias, M. R., Carneiro, J. C., Schneider, H., de Alencar, T. B., Calouro, A. M., Dalponte, J. C., de Souza Mattos, F., Ferrari, S. F., Buss, G., & de Azevedo, R. B. (2019). A new species of Titi monkey, Plecturocebus Byrne et al., 2016 (Primates, Pitheciidae), from southwestern Amazonia, Brazil. *Primate Conservation, 33*, 21–35.

Hallmann, C. A., Sorg, M., Jongejans, E., Siepel, H., Hofland, N., Schwan, H., Stenmans, W., Müller, A., Sumser, H., Hörren, T., & Goulson, D. (2017). More than 75 percent decline over 27 years in total flying insect biomass in protected areas. *PLoS One, 12*(10), e0185809.

Hoffmann, M., Duckworth, J. W., Holmes, K., Mallon, D. P., Rodrigues, A. S., & Stuart, S. N. (2015). The difference conservation makes to extinction risk of the world's ungulates. *Conservation Biology, 29*(5), 1303–1313.

Howard, B.C. 2019. First mammal species recognized as extinct due to climate change. National Geographic. Published February 20, 2019, Accessed 25 February, 2019. https://news.nationalgeographic.com/2016/06/first-mammal-extinct-climate-change-bramble-cay-melomys/.

Johnson, C., Cogger, H., Dickman, C., & Ford, H. (2007). *Impacts of landclearing: The impacts of approved clearing of native vegetation on Australian wildlife in New South Wales*. Sydney: WWF Australia report.

Ketterson, E. D. (1979). Aggressive behavior in wintering dark-eyed juncos: Determinants of dominance and their possible relation to geographic variation in sex ratio. *The Wilson Bulletin*, 371–383.

Krech, S. (1999). *The ecological Indian: Myth and history*. WW Norton & Company.

LaPointe, D. A. (2006). Feral pigs, introduced mosquitoes, and the decline of Hawaii's native birds. USGS Report no. 3029.

3

Lewis, S. M., Wong, C. H., Owens, A., Fallon, C., Jepsen, S., Thancharoen, A., Wu, C., De Cock, R., Novák, M., López-Palafox, T., & Khoo, V. (2020). A global perspective on firefly extinction threats. *Bioscience, 70*(2), 157–167.

Margalida, A., Ogada, D., & Botha, A. (2019). Protect African vultures from poison. *Science, 365*(6458), 1089–1090. https://doi.org/10.1126/science.aay7945.

Mateo-Tomás, P., & López-Bao, J. V. (2020). Poisoning poached megafauna can boost trade in African vultures. *Biological Conservation, 241*, 108389.

Mindo Cloudforest Foundation (n.d.). Mindocloudforest.org.

Myers, R. A., Baum, J. K., Shepherd, T. D., Powers, S. P., & Peterson, C. H. (2007). Cascading effects of the loss of apex predatory sharks from a coastal ocean. *Science, 315*(5820), 1846–1850.

Nicholls, H. (2015). Mysterious die-off sparks race to save Saiga antelope. *Nature News.*

Owens, B. (2017). Behind New Zealand's wild plan to purge all pests. *Nature News, 541*(7636), 148.

Pardo, D., Forcada, J., Wood, A. G., Tuck, G. N., Ireland, L., Pradel, R., Croxall, J. P., & Phillips, R. A. (2017). Additive effects of climate and fisheries drive ongoing declines in multiple albatross species. *Proceedings of the National Academy of Sciences, 114*(50), E10829–E10837.

Patel, R., & Moore, J. W. (2017). *A history of the world in seven cheap things: A guide to capitalism, nature, and the future of the planet*. Univ of California Press.

Pimm, S. L., Russell, G. J., Gittleman, J. L., & Brooks, T. M. (1995). The future of biodiversity. *Science, 269*(5222), 347–350.

Prakash, V., Bishwakarma, M. C., Chaudhary, A., Cuthbert, R., Dave, R., Kulkarni, M., Kumar, S., Paudel, K., Ranade, S., Shringarpure, R., & Green, R. E. (2012). The population decline of Gyps vultures in India and Nepal has slowed since veterinary use of diclofenac was banned. *PLoS One, 7*(11), e49118.

Oaks, J. L., Gilbert, M., Virani, M. Z., Watson, R. T., Meteyer, C. U., Rideout, B. A., Shivaprasad, H. L., Ahmed, S., Chaudhry, M. J. I., Arshad, M., & Mahmood, S. (2004). Diclofenac residues as the cause of vulture population decline in Pakistan. *Nature, 427*(6975), 630.

Ogada, D., Botha, A., & Shaw, P. (2016). Ivory poachers and poison: Drivers of Africa's declining vulture populations. *Oryx, 50*(4), 593–596.

Rasner, C. A., Yeh, P., Eggert, L. S., Hunt, K. E., Woodruff, D. S., & Price, T. D. (2004). Genetic and morphological evolution following a founder event in the dark-eyed junco, Junco hyemalis thurberi. *Molecular Ecology, 13*(3), 671–681.

Readfearn, G., & Hodgens, P. (2020). Silent death': Australia's bushfires push countless species to extinction. *The Guardian*, 2020. https://www.theguardian.com/environment/2020/jan/04/ecologists-warn-silent-death-australia-bushfires-endangered-species-extinction?CMP=Share_iOSApp_Other.

Robertson, G., Moreno, C., Arata, J. A., Candy, S. G., Lawton, K., Valencia, J., Wienecke, B., Kirkwood, R., Taylor, P., & Suazo, C. G. (2014). Black-browed albatross numbers in Chile increase in response to reduced mortality in fisheries. *Biological Conservation, 169*, 319–333.

Rothman, D. H. (2017). Thresholds of catastrophe in the earth system. *Science Advances, 3*(9), e1700906.

Rosenberg, K., et al. (2019). Decline of the north American avifauna. *Science*. https://doi.org/10.1126/science.aaw1313.

Ruppenthal, A. (2019). Fearing the 'insect apocalypse'? Renowned entomologist says 'get Rid of your Lawn'. *WTTW PBS*.

Russell, J. C., Innes, J. G., Brown, P. H., & Byrom, A. E. (2015). Predator-free New Zealand: Conservation country. *Bioscience, 65*(5), 520–525.

Saidu, Y., & Buij, R. (2013). Traditional medicine trade in vulture parts in northern Nigeria. *Vulture News, 65*(1), 4–14.

Soulé, M. E. (1987). *Viable populations for conservation*. Cambridge university press.

Spitzen-van der Sluijs, A., Spikmans, F., Bosman, W., de Zeeuw, M., van der Meij, T., Goverse, E., Kik, M., Pasmans, F., & Martel, A. (2013). Rapid enigmatic decline drives the fire salamander (Salamandra salamandra) to the edge of extinction in the Netherlands. *Amphibia-Reptilia, 34*(2), 233–239.

Stuart, S. N., Chanson, J. S., Cox, N. A., Young, B. E., Rodrigues, A. S., Fischman, D. L., & Waller, R. W. (2004). Status and trends of amphibian declines and extinctions worldwide. *Science, 306*(5702), 1783–1786.

Sugden, A. M. (2002). The decline of the albatross. *Science, 295*(5553), 235. https://doi.org/10.1126/science.295.5553.235g.

Thompson D. (2012). The spectacular rise and fall of us whaling: An innovation story. *The Atlantic. [Electronic version]*. Retrieved 11 July, p. 2016.

Tuck, G. N., Polacheck, T., Croxall, J. P., & Weimerskirch, H. (2001). Modelling the impact of fishery by-catches on albatross populations. *Journal of Applied Ecology, 38*(6), 1182–1196.

U.S. Fish and Wildlife Service. (2008). Short-tailed Albatross Recovery Plan.

Vance, E. 2020. How juncos changed their migration, behavior, and plumage in a matter of decades. Living bird magazine, winter issue, the Cornell lab of ornithology.

Vulture Apocalypse (Nat Geog). (2011). (Accessed on April 13, 2018, URL: https://www.youtube.com/watch?v=6a49CATa3f0).

Welford, M. R., & Yarbrough, R. A. (2015). Serendipitous conservation: Impacts of oil pipeline construction in rural northwestern Ecuador. *The Extractive Industries and Society, 2*(4), 766–774.

Whitehorn, P. R., O'connor, S., Wackers, F. L., & Goulson, D. (2012). Neonicotinoid pesticide reduces bumble bee colony growth and queen production. *Science*, 1215025.

Winkler, D. W., Gandoy, F. A., Areta, J. I., Iliff, M. J., Rakhimberdiev, E., Kardynal, K. J., & Hobson, K. A. (2017). Long-distance range expansion and rapid adjustment of migration in a newly established population of barn swallows breeding in Argentina. *Current Biology, 27*(7), 1080–1084.

Wittemyer, G., Northrup, J. M., Blanc, J., Douglas-Hamilton, I., Omondi, P., & Burnham, K. P. (2014). Illegal killing for ivory drives global decline in African elephants. *Proceedings of the National Academy of Sciences, 111*(36), 13117–13121.

Woinarski, J. C. Z., Braby, M. F., Burbidge, A. A., Coates, D., Garnett, S. T., Fensham, R. J., Legge, S. M., McKenzie, N. L., Silcock, J. L., & Murphy, B. P. (2019). Reading the black book: The number, timing, distribution and causes of listed extinctions in Australia. *Biological Conservation, 239*, 108261.

Worm, B., Davis, B., Kettemer, L., Ward-Paige, C. A., Chapman, D., Heithaus, M. R., Kessel, S. T., & Gruber, S. H. (2013). Global catches, exploitation rates, and rebuilding options for sharks. *Marine Policy, 40*, 194–204.

4

Thresholds

Contents

M. R. Welford, R. A. Yarbrough, *Human-Environment Interactions*,
https://doi.org/10.1007/978-3-030-56032-4_4

Learning Goals

After reading this chapter, you will be able to:

- Evaluate whether human activity and pollution have exceeded the capability of the planet to absorb these actions.
- Identify areas in which we have exceeded critical environmental thresholds.
- Assess the threshold cost and consequences for continuing oil use.
- Evaluate effects on global warming, agriculture and food consumption, travel and innovation, multinationals and nations when we surpass the world's oil cost threshold.

4.1 Thresholds and Scales

In this chapter, we explore the notion that changes in human activities, resource availability, and environment processes and patterns occur after crossing a threshold or divide. Typically, human-environmental process thresholds involve exceeding some kind of environmental carrying capacity that appears to depend spatially on population, affluence and technological (IPAT—Impact = population * affluence * technology). These thresholds can be one-directional as in the formation of arroyos, or bidirectional in bark beetle population eruptions, or change entire ecosystems as in the near extinction of India's vulture populations. Establishing the cause-and-effect for most human-landscape-atmosphere process relationships is difficult. Typically, environmental-process thresholds are hard to identify prior to threshold exceedance. Moreover, to add to the uncertainty, thresholds that separate different process regimes can be gradual or abrupt. In addition, most environmental systems are highly buffered across space and time—in other words, multiple different process thresholds must be exceeded within a system before environmental change is observable, and when change is observable, it frequently appears chaotic and difficult to ascribe to single causes-and-effects. Nevertheless, global environmental degradation might have reached or in many cases exceeded their critical thresholds, such that the extent and severity of the impacts may prove irreversible.

All thresholds are scale-dependent and spatial in nature and appear nonlinear. In other words, they are complex in nature, hard to identify before a threshold is exceeded, and vary across the landscape. Thresholds can be sensitive or insensitive to significant changes in inputs. The acceleration, or graphic hockey-stick, in the global temperatures witnessed over the last decades, that is, the five warmest years on record have occurred since 2014, suggests that global climate change has recently exceeded a critical threshold.

Floods seem to be threshold-dependent. The Cedar and Upper Mississippi River Valleys have recently recorded three 500-year floods in July 1993, July 1999, and June 2008! These events followed large late spring snowfalls, wet springs, and early summers that appear to be associated with a weakening of the polar front and the increased penetration of tropical air masses into the upper Midwest.

4.2 Landscape Sensitivity and Complex Responses

Abrupt or gradual changes to the natural environment all involve some form of threshold exceedance. Jonathan Phillips suggests they are typically characterized by the "four Rs." How quickly the system *responds* to a change in input and how quickly it adjusts to a sustained change in the system or inputs, the *resistance* of a system to drivers of change, while the degree of change is mitigated by the fragility or *resilience* of the environment to change, and *recursion* of a system to either undergo positive or negative feedback.

This can be encapsulated within the transient form ratio (TF ratio) developed to evaluate change within landscapes.

- TF > 1: the mean recurrence time of events capable of producing change is shorter than the time taken for the system (or component of the system) to recover or equilibrate, there is likely to be a poor correspondence between process agents and resulting forms (be they landforms, ecosystems); that is, forms will be predominantly transient. In this case, the system can be considered fragile.
- TF < 1: the system has the potential to adjust to new conditions before the next major disturbance so that characteristic forms will tend to prevail after the initial recovery period, leading to more reliable process-response relationships. In this case, the system can be considered resilient.
- Brunsden and Thornes (1979)

Abrupt, episodic shocks to the landscape are typically followed by a return to the preexisting form or pattern, or TF < 1, whereas perturbations that are ramped and continue to increase in magnitude can result in permanent, long-lasting changes to the landscape and exhibit TF> 1. The introduction of fire to the East African savannah by humans 2–3 million years ago is an excellent example of a ramped input that opened up the forest, creating broad, open grasslands. A more nuanced perspective suggested by Phillips and Van Dyke is that landscapes can exhibit high resistance and resilience to change (e.g., TF < 1), rapid relaxation times, and stable recursive feedbacks that mitigate the effects of disturbance. In other words, irrespective of the magnitude of the disturbance, be it human or natural, for example, hurricane-induced flooding, some landscapes quickly adjust to the disturbance as feedbacks minimize the effect of the disturbance, such as a river quickly eroding, transporting, and moving sediments downstream after a flood.

▶ Example

A good example is the Big Thompson Flood of July 31, 1976, where 12 to 14 inches of rain fell in four hours in its upper drainage near Estes Park. By 9 pm, the creek had risen from 18 inches to 20 feet, killing 143 people. Today, other than a stone memorial and the debris fan, little evidence of the catastrophe remains in the canyon. River flow has remodeled its channel to its pre-flood conditions such that the forms created by the flood are obliterated. ◀

In contrast, low resistance and resilience (e.g., TF > 1), slow relaxation, and unstable feedbacks can, according to Phillips and Van Dyke, create large impacts that remain on the landscape for thousands of years. We might be witnessing just such a phenomenon with the persistent shrinkage of the North Polar sea ice.

Recursive and non-recursive landscape feedback loops, mentioned in the preceding section, were first identified by Schumm in the 1960s. Arroyo and terrace formation in the American West illustrate these feedback loops. Arroyos are gully-like landforms that carry water flow only intermittently and were first observed across the American West in the late 1890s. Typically, arroyos exhibit steep or vertical walls in cohesive, fine sediments and have flat and generally sandy floors. They tend to desiccate the surrounding soils as water tables drop to the lower river floor. They coincided with the fencing of the American West that constrained the formerly open range cattle. Cattle over-grazed and locally compacted the surrounding landscape, reducing rainfall infiltration capacities that led to higher volumes of overland flow and higher in-stream flows that increased flow shear stress in the channel perimeter, leading to channel scour and channel entrenchment.

Among terraces, a single perturbation to the river system (where TF > 1) triggered by climate change, grazing, compaction of the drainage surface upslope from the terrace, base-level change in the form of local or regional uplift or local lake drainage or sea-level drop might precipitate channel erosion, sediment transport and removal, and down-cutting of the streambed and abandonment of the floodplain as a nick-point migrates upstream. Simply put, terraces are abandoned floodplains. However, sediment transport can overwhelm stream capacity leading downstream to deposition and aggradation and the formation of lower alluvial stream terraces within the original incised floodplain. Wolman and Schumm observed just such a phenomenon in Douglas Creek in 1978. So, within the channel, riverbeds are eroding at one point at and immediately below a nick-point but aggrading further downstream.

However, streams and rivers are not disconnected from their surrounding watershed slopes. To clarify, a watershed is all the landscape—its soil, its subsurface rocks, its slopes, hills, floodplains, and terraces that contribute water to a stream or river. Rain falling on a watershed moves over slopes, but mostly through the soil and rock, to where it intersects stream or river courses and then flows into the stream channel. If during heavy rainfall events, the instantaneous infiltration capacity threshold of the watershed surface materials—rock or sediment or soil—is exceeded, then water flows downslope across surface slopes to streams and rivers. Sheets of overland flow or surface wash are not commonly observed; rather, surface wash becomes concentrated in the form of rills as water is diverted around obstacles. These obstacles can vary from the size of individual sand grains to plant and tree stems and roots. Once rills form and flow, they can entrain larger particles than surface flow. Rills are ephemeral and short-lived and new rill networks frequently develop after each rain storm. Rills are typically less than 0.5 m in depth, do not extend to the head of slope, and run parallel to each other on fresh surfaces, developing a branching network as divides between rills are broken down by lateral migration or rain splash erosion.

Rills develop on unstable, typically weak vegetated or exposed soil slope surfaces that have been burnt or overgrazed by animals and whose surface is compacted or experienced either significant increases or decreases in rainfall due to local or regional changes in climate, or suffered human-induced deforestation, or suffered the spread of cultivation or urban sprawl.

If unstable slope surface conditions persist, rills can on the lower slopes become gullies. Gullies exceed 0.5 m in width and depth and may be in-excess of 10 m across. They are, in fact, enlarged rills and become permanent landscape features. Some gullies extend to permanent stream channels. They are steep-sided, V-shaped, and flat-floored channels. They represent major sites of sediment storage on slopes at the end of major rainfall events.

Gullies are the principal conduit of slope sediment to streams, but their size is controlled by the length of their contributing slopes or catchment area. Gullies result in accelerated erosion from slopes and increase the speed and volume of water reaching streams (Fig. 4.1).

Gullies increase flood runoff but decrease the time to peak runoff, leading to more flash floods. However, in the longer term, gullies decrease the capacity of channels to carry runoff as gullies contribute too much sediment to streams and rivers, thereby choking channels.

Adrian Harvey established over a 20-period that within gully, interactions are seasonal, whereas those related to the coupling of slope/streams, where debris accumulation in the form of colluvial cones and fans and their removal by floods and basal scour by streams, produce a cyclicity over 2–5 years. He also established

Fig. 4.1 Gullies in North-central Idaho (co-author Mark Welford and Scott Morris for scale) @ Mark Welford

Fig. 4.2 Active and inactive gullies where reforestation in the channel slopes is stabilizing the lower portion of the basin, while several upper slopes are still actively eroding (unknown source)

that over decades, progressive changes in gully morphology and revegetation of once unstable contributing valley slopes may lead to negative feedback mechanisms that stabilize the gully, resulting in a finite age of gullies and an upper size limit for modern gully development (Fig. 4.2).

▸ Example

Hydraulic gold mining with its associated use of mercury to trap the placer gold among the South Fork of the American River in California is an excellent example of the aforementioned process and has left an environmental disaster. Over 7500 tons of mercury was used between 1849 and 1884 to trap gold in hydraulic mining sluices. Allan James estimates the hydraulic mining also caused in excess of 13 billion tons of sediment to be eroded within the watershed. Banning hydraulic mining in 1884 did not immediately mitigate the effects of mining. The 13 billion tons of sediment continue to episodically move through the watershed. In fact, in the lower Bear Basin within the mining districts, half of the mining sediment remains. These deposits cover 50 km^2 to depths of 2 to 3 meters. Initially, the riverbed of the South Fork aggraded, with sediment washing over banks and submerging adjacent valley sides and forest due to increased frequency of flood events. As a result, spatially disjunct slugs of riverbed sediment have been mobilized where degradation of the riverbed has occurred, as erosional nick-points have moved upriver. Courser gravel-sized sediment has been deposited as alluvial fans upon the Central Valley, east of Sacramento, creating dry xeric soils highly suitable for citrus farms. Further downriver, more mobile finer silt and clay-sized particles contaminated with mercury from hydraulic mining are filling San Francisco Bay. Bioaccumulation of mercury has been observed among the bay's *Macoma balthica* clams. ◀

► **Example**

The recent Ok Tedi environmental disaster in Papua New Guinea mirrors the processes of aggradation, degradation, and contamination observed in the American River. Between 1984 and 2013, over 260 million tons of mine waste were discharged into the Ok Tedi and Fly Rivers. Just below the mine, the river aggraded 10 meters, while a thick toxic gray sludge continues to migrate downstream. Over 1300 km^2 of flood plain has been inundated due to the aggradation of the streambed contaminating taro, bananas, and sago palm fields. While a further 1600 km^2 of forest has died. Approximately 50,000 people have suffered economic or medical hardships. Although a US$28.6 million out-of-court settlement was reached, few have benefited from the settlement, while the mining company was exempted from further fines. ◀

Recent work does suggest that geomorphic systems are responding to recent rapid climate change. Andrew Goudie suggests rates of landscape erosion will increase between 25–50% by 2100, particularly in cold, tropical, and arid drainage systems, while deltas will have to respond to sea level rise and increased sedimentation due to the transfer of these landscape erosion products. James Knox argues that we are already observing large increases in flood magnitude due to modest changes in climate. An intense snow ablation event witnessed in France in 2003 suggests that if global warming is sustained, sediment transport from alpine regions will increase significantly and likely compromise the flood retention and water storage capacities of many of the region's dams.

Knorr and coauthors suggested long-term global warming will lead to significant soil carbon escaping into the atmosphere; according to Sazonova and coauthors, this prediction is coming true as the arctic tundra is venting tens of thousands of teragrams of methane into the atmosphere.

4.3 Carrying Capacity Exceedance

Although change can be minor, cumulative, or abrupt within the natural environment, any change can be detrimental to societal health. Jared Diamond in his book *Collapse* suggested that environmental damage, climate change, hostile neighbors, and the loss or lack of friendly trade partners could prove instrumental to a society collapsing. In contrast, a society's responses to its environmental problems have, it appears, always proved significant to a society's ability to either stave off collapse or collapse.

Lastly, in many ways, Easter Island foretells the future of the earth, we have no intergalactic trading partners, and we are knowingly and willingly engaged in local, regional, and global environmental destruction because we are economically, politically, and culturally too inflexible to change our consumption habits.

Easter Island

In perhaps the most famous example of societal collapse and accompanying localized faunal and floral extinctions, all seabirds, indigenous birds, and palms were expiated from Easter Island by 1450 CE, yet Easter Island was first occupied in 900 CE by the Rapa Nui civilization. In just 550 years, the Rapa Nui had stripped all viable resources from the island, including much of the island's soil as soil erosion intensified when the palm forests were cut down. The pace of decline mirrored the destruction of the island's soils. According to Mieth and Bork, prior to 1280 CE, sustainable agriculture and agroforestry limited soil erosion; however, after 1300 CE, soil erosion accelerated, burying many settlements and ceremonial sites in rill and gully sediments. Today, many of these gullies are still active, illustrating the fragility of the island's slopes and soils.

In the defense of the Rapa Nui, the island's isolation and subtropical/temperate climate were not conducive to large-scale human occupation. Easter Island also lacks fringing coral reefs or a lagoon, which meant that fish and shellfish contributed little to the diet of the Rapa Nui. Instead, protein came from seabirds, landbirds, porpoises, and introduced chickens—no pigs or goats made the long sea voyage. This was coupled with the lack of trading partners, as Easter Island is the most isolated island in the Pacific: lacking trade partners meant the Islanders were on their own.

Instead, the Rapa Nui concentrated on transforming the island through deforestation to gardens for cultivating potatoes, yams, taro, bananas, and sugarcane. For building monumental platforms and statues, palm logs were used to transport the increasingly large moai from the highlands to the coast. As a result of both enterprises, Easter Island was completely deforested by 1450 CE. The deforestation also reduced the ability of the Islanders to utilize the limited marine resources. When the first Europeans ventured onto Easter Island, the few remaining canoes were small and leaked, and they were incapable of sustained ocean travel. The pace of deforestation also far outstripped the island's natural ability to reforest because of three factors. First, Easter Island is cool and dry, nearly temperate in nature. Second, Easter Island lies far from the fertile Pacific volcanic dust cloud. As a result, Easter Island's soil is poor and prone to erosion. And third, seed-recruitment failed across the island as the abundant presence of rats ensured that most palm seeds (recovered from the island during this time of environmental stress at the peak of moai production) show evidence of rat tooth marks. By 1722 when Easter Island was first visited by Europeans, the Rapa Nui, the palm forest, and all the indigenous land birds were extinct, and most of the 30 or more breeding seabirds were no longer breeding on the main island.

IPAT: local, regional, and global

Unintended consequences of IPAT are widespread, but one of the most surprising is the connection between the near extinction of several Indian vulture species and the explosion in rabies cases across India.

▶ Example

In the last few decades, India's economy has boomed, and today India is one of the fastest growing major economies on earth. As the number of elite and affluent households increased, so did the efforts to relieve suffering among India's sacred cows. Beginning in the mid-1980s, the painkiller and anti-inflammatory drug, diclofenac, was increasingly given to cattle, even sacred street cattle. However, among vultures, diclofenac causes visceral gout, a deadly disease that ultimately killed 98% of all lowland and foothill vultures in India. At the same time that vultures vanished from India's cattle carcass dumps, the number of street dogs increased by some 7 million.

Cattle carcass dumps are sites where cattle carcasses are dumped, as only 4% of all cows which are considered sacred are eaten by Indians. Before the widespread use of diclofenac, up to 15,000 vultures could be observed at carcass dumps in just New Delhi. However, once the vultures began to decline, street dogs gained more access to cattle carcasses at cattle dumps. This shifted the balance of rabies infections across India. Vultures with their highly acidic stomachs are dead-ends to nearly all the pathogens they consume, including rabies. So as vultures decreased and dogs increased, the incidence of rabies increased. However, where leopards are present, the numbers of rabies deaths have not risen so steeply, as leopards have targeted street dogs but this has meant more attacks on humans. In fact, the first attack on a human by a leopard near Mumbai in an adjacent national park was in 1986. Deforestation and human encroachment into parks was thought to explain leopard attacks. But leopard scat analysis suggests leopards were targeting street dogs and human attacks were incidental. Throughout the 1990s, India suffered more and more leopard attacks on humans, just as India saw more and more street dogs and more and more rabies cases. Since the early 1990s, India has seen an extra 50,000 cases of rabies, and today, in India, there are 36% of all worldwide rabies cases and in excess of 20,000 human deaths per year, at a cost of 750 million rupees per year. The Indian government is funding vulture breeding and reintroducing efforts across India to eliminate rabies. Sky-burials by Mumbai's 3000-year-old Zoroastrian Parsi community at their Tower of Silence have had to be curtailed as the vultures have disappeared. Although raven and kite numbers have increased, these birds are not efficient at consuming human flesh, and as a result the bodies rot and their smell is increasingly disturbing local residents. ◀

▪▪ Sprawl and Species-Area Relationships

The single greatest threat to biodiversity in the USA is suburban sprawl and the expansion of the wildland-urban interface or boundary. Since European settlement of NA, 27 different types of natural communities have declined by 98%. The major single cause of biodiversity loss according to Ehrlich is habitat loss and degradation. Destruction of previously intact ecosystems results in a loss of habitat for multitudes of species and breaks down an ecosystem's ability to function. In fact, around the world, urban populations are growing faster than rural populations. In the USA, Forman and Deblinger found that bird densities are significantly reduced within 1000 meters of roads due to the fragmentation of forest and the increase in brood parasitism by cowbirds, as cowbirds invade the fragmented landscape, while road densities lead to significant loss of wetland-dependent amphibians, lizards, and bird species, although Findlay and Bourdages found that these losses lag a decade behind any road development. Urban sprawl, defined as low-density development of natural areas outside of cities and towns and its associated wildland-urban interface or boundary that directly elevates fire risk, according to Syphard and coauthors, is responsible for threatening 188 of California's 286 endangered species. Population density and the distance to the wildland-urban interface appear to determine the spatial pattern of fire-risk, with the highest fre-

4

quency of fire and burnt areas occurring at intermediate levels of human activity, but once population densities exceed critical thresholds, fire frequency declines as less fuel is present on the suburban landscape. Yet effective planning in rapidly urbanizing landscape of Eastern Pennsylvania has diminished by half the proportion of habitats lost.

Sprawl also exacerbates air and water pollution that reduces both biodiversity and critical ecosystem services, where ecosystem services are defined as natural capital assetsthat supply life-support services of tremendous value to the landscape and humans. These services include, but are not limited to, services that purify air and water, mitigate floods and droughts, and detoxify and decompose wastes. For instance, new construction accelerates soil erosion as forest, shrubs, and grasses are cleared for development. Eroded soil is transported downslope into streams through surface flow, and rill and gully erosion (see ◘ Figs. 4.1, 4.2, and 4.3), resulting in increased stream siltation which can reduce the life of dams. Deforestation immediately adjacent to streams can lead to rising stream temperatures and stream eutrophication, stream bank erosion, stream entrenchment and arroyo development, and increasing flash floods as surface infiltration capacities decline. As natural ecosystems shrink, there is less natural capacity to filter pollutants and detoxify waters and less capacity to cycle nutrients and compost organic wastes. As sprawl increases, species and ecosystem services decrease. Small, fragmented habitats hold less than large, single habitat. In Chicago, "unassociated vegetation," which represents fragmented, degraded patches of nonnative woody and grassy vegetation, increased from 7% to 22% of land use/cover as sprawl expanded north, west, and south of the urban core. Some animals benefit from sprawl such as dietary generalists, previously cliff-dwelling birds, and granivores, in other words, those animals that eat seeds as a main part of their diet. Highly disturbed environments such as urban and suburban gardens and brown sites are dominated by weeds that produce large numbers of easily distributed seeds.

Sprawl is not just a local issue; nonhuman creatures and non-domesticated plants are running out of space even on continents. And what space is left is highly fragmented. At the crux of the issue of space is the species-area relationship. Simply put, the larger the area, the greater the number of individuals and species an area can hold. So smaller islands or smaller continental fragments or smaller patches of habitat have few species, because these small islands or patches of habitat have less space and less variation in environments, while larger islands or larger patches of habitat have more species, more space, and typically a larger variety of habitats which can support a richer variety of ecosystems. This can occur among true islands surrounded by water, or mountains surrounded by lowlands, or patches of habitat surrounded by urban sprawl or arable land. In the case of mountains, in the American West, several mountain ranges lie surrounded by inhospitable deserts. These mountain-islands or sky islands do exhibit species-area threshold relationships (Johnson 1975), although they display higher species diversity and comparatively low endemism compared to true islands, especially when compared to oceanic islands (McLaughlin 1994). Yellowstone NP is surrounded by a low-

Fig. 4.3 The Little Grand Canyon of the South or Providence Canyon, Georgia, illustrates a mega gully initiated in the late 1920s. @ Mark Welford

density urban/suburban/arable sprawl that restricts migration of mammals into the park. For all intent and proposes, Yellowstone is a sky-island! It exhibits a lower mammalian and amphibian diversity than other more connected habitat patches in the Rocky Mountains. However, the reintroduction of the wolf in 1995 is changing this dynamic.

► Species-Area Threshold Extinctions

Barro Colorado Island (BCI) in Panama was created by the damming of the Chagres River in 1913 to form Gatun Lake that provides water to both sides of the Panama Canal. Since its isolation, the understory of BCI has opened up as tree seedling recruitment has slowed. Two large-seeded, mature-phase rain forest trees *Dipteryx panamensis* and *Gustavia superba* are not regenerating. Their seeds have ten times greater likelihood of surviving on mainland than on BCI. About 50-60 bird species have disappeared from BCI. Among undergrowth forest bird species, rarity is not a good predictor of extinction on BCI; instead, it appears that birds that utilize patchy or seasonally variable resources, be they insectivores, frugivores, or nectarivores, are more prone to extinctions (Karr 1982). In other words, there are insufficient patchily or seasonally variable resources in the small area of BCI to support these bird populations. Many ground-nesting birds also disappeared as coatimundis and opossums robbed nests. In fact, 96% of ground-nesting attempts fail on BCI, but only 4% on mainland. It appears that as cats (jaguar) left BCI (only one has ever been observed on BCI in 1987), so coatimundi and opossum numbers exploded, killing off ground-nesting species. The following table illustrates the impact of jaguars disappearing from BCI (Cocha Cashu is on the mainland).

	Number of individuals per km^2	
	Cocha Cashu	BCI
Opossum	20	47
Armadillo	4	53
Agouti	5	100
Paca	4	40

The tropic cascade triggered by the disappearance of jaguars from the BCI forest has led to the local extinction of many bird species, while seed predators such as the agoutis, peccaries, paca, and rats have expanded and increased seed-predation rates within BCI. Asquith and coauthors suggest that the forest of BCI is in fact dying. In a sense, BCI is a microcosm of the earth, both terrestrial and marine apex predators across the globe are in rapid decline, and the resultant tropic cascades will be significant, detrimental, and unpredictable! Rather sadly, the Smithsonian Tropical Research Institute established on BCI is a witness to this ecosystem death, not something the founders expected. ◄

▶ Species-Area and Hunting Thresholds

Sadly, the highly endangered helmeted hornbill (*Rhinoplax vigil*) illustrates the problems of rampant deforestation, hunting, and minimum viable populations. A minimum viable population is a genetically isolated population of at least 50 pairs that have at least a 95% chance of surviving extinction over the next 100 years despite demographic, environmental, genetic stochasticity, and natural catastrophes. Phillipps suggests helmeted hornbills eat large figs exclusively; as a result, each pair needs approximately 7.7 km^2 of territory to breed successfully and 50 pairs require 385 km^2 of intact lowland rainforest. However, in Borneo, an area of only 743,380 km^2, only two intact areas of contiguous lowland forest in excess of the minimum threshold area, remains: one at Danum Valley Conservation Area of 438 km^2, and one at Tabin Wildlife Reserve of 1225 km^2. However, Tabin Wildlife Reserve has an intact, unlogged core area that is significantly smaller than 1225 km^2. These populations are protected, but other surviving helmeted hornbill populations outside of these two parks also suffer significant hunting pressures. Today, helmeted hornbill's casque, its red helmet-like solid protrusion on its upper mandible, is highly sought after by Chinese ivory carvers. In Laos, helmeted hornbill heads go for $3700. ◀

Even under pre-agricultural, pre-urban conditions where humans existed at quite low densities, we have wielded a big deadly club! Although the explanations for megafauna extinctions in the post-glacial environment are many and varied, one consistent theme is threshold related—the notion of overkill or human hunter-facilitated extinction. In the North America, by 11,700 years BP, some 70% of the megafauna and 45% of the predatory bird genera including a giant vulture *Teratornic incredibilis* had gone extinct. In a very interesting experiment, John Alroy simulated the impact of growing human populations in North America on megafauna. His simulations assumed megafauna and humans had slow population growth rates, that hunting was random, and hunting effects were low; in other words, few were killed each year. Nevertheless, varied simulations had a 78% success rate in predicting a rapid megafauna extinction in and around 11,700 years BP. Although humans were not a plague on NA by this point, the low reproductive rate of megafauna coupled with slowly rising human populations was sufficient to push megafauna (those animals over 100 kilograms) over a critical population threshold and on to extinction shortly after the end of the Late Wisconsin Ice Age. It probably did not help that this was a time of significant climate and environment change precipitated by the end of the Pleistocene Ice Age and what appears to be a 31-km-wide meteorite impact during the Younger Dryas around 12,900 BP under what is now Hiawatha Glacier in NW Greenland. According to Paul Voosen and Kurt Kjær, shocked quartz, iridium (both evidence of impacts), and platinum have been found downstream from the Hiawatha glacier and in Greenland ice cores from this time, but from a 1000 km south of the impact site. Firestone and coauthors suggest this event contributed to the NA megafauna extinction and disappearance of the Clovis native American culture.

Even abundant, widespread species can and do go extinct. *Ectopostes migratorius*, the passenger pigeon was quite possibly the most abundant bird or non-insect animal to go extinct. In 1813, James J. Audubon observed over three days a mega-flock of an estimated 1.11 billion birds. But its extinction is complicated. Certainly, the pigeon was hunted for food. One boxcar near Chicago was discovered containing 38,000 dead birds, and over 5 months in Petoskey, Michigan, in 1878, 7.5 million birds were killed, while the American chestnut tree (*Castanea dentata*), one its principal food sources, was decimated by the introduction of a chestnut blight. Recent genetic work and modeling by Chih-Ming Hung and coauthors suggest that the passenger pigeon was an "outbreak species" with a small genetically effective population—the product of repeated boom-and-bust cycles in its population due to its diet specialization on acorns, chestnuts, and beechnuts that vary between 12-fold and 136-fold in seasonal production. With the loss of the American chestnuts and subsequent overconsumption of acorns and beechnuts and overhunting, the population of the passenger pigeon crashed in the late 1890s. The huge flocks of passenger pigeon were selected for in order to locate sufficient food because its food underwent such massive spatial variability in seasonal production within the Eastern USA. However, David Quammen believes that once flock sizes decreased below a threshold size, possibly 500,000 individuals, the flocks were unable to locate food and the pigeon spiraled into extinction.

4.4 Deforestation, Colonization, Emergence of New Diseases, and Reemergence of Known Diseases

The emergence of new diseases such as Ebola, HIV-1, Marburg virus, Lasa fever, or Hanta virus, or the reemergence of long-known diseases such as tuberculosis, is the product of various environmental factors working together to create conditions that support transmission of a pathogen to humans, the maintenance of the pathogen with a spatial restricted human population, and then the pathogens' transmission to a broader spatial scale and a larger human population. Stephen Morse suggests that among the principle factors that contribute to the emergence of infectious pathogens are human-induced changes in ecological systems; changes in human demographics and behavior; increasing human mobility, industrialization, microbial adaptation, and change; and the failure of public health services (▫ Table 4.1).

What is frightening is that these factors are not abating; rather, they are getting more severe as human populations continue to grow, consume national resources, and pollute and alter natural ecosystems, making them simpler and more prone to catastrophic fluctuations.

Table 4.1 Factors that lead to the emergence of pathogens (altered from Morse 2001).

Factor	Examples	Diseases
Environmental and Land-use changes	Agriculture; dams; deforestation; rural sprawl; introductions of alien species; climate change; climatic fluctuation; bush-meat consumption	Schistosomiasis (dams); Ebola and Marburg Hemorrhagic Fever (deforestation, bush-meat); Lasa Fever (rural sprawl); Rift Valley Fever (dams, irrigation); Hanta Virus (weather anomalies that increase contact with rodent hosts); HIV-1 (bush-meat)
Human demographic changes	Societal events: Population growth and migration (movement from rural areas to cities); war or civil conflict; urban decay; sexual behavior; intravenous drug use; use of high-density facilities	Dengue; HIV (colonialism, inoculations, blood product trade); Cholera (Vibrio cholerae O1, serotype Ogawa, biotype El Tor introduced into Haiti by Nepalese UN troops, 2010)
Human Mobility	Worldwide movement of goods and people; air travel	Malaria; SARS (airlines and super-spreaders); COVID-19; Pneumonic Plague (humans, 1347-1815); Bubonic Plague (rats via shipping, ~1870s to present-day)
Industrialization	Globalization of food supplies; changes in food processing and packaging; organ or tissue transplantation; drugs causing immunosuppression; widespread use of antibiotics and anti-inflammatory drugs	Hemolytic uremic syndrome (*E.coli* contamination of hamburger meat), bovine spongiform encephalopathy (rendering of sheep infected with scabies for cattle protein supplements); transfusion-associated hepatitis (hepatitis B, C), opportunistic infections in immunosuppressed patients, Creutzfeldt-Jakob disease from contaminated batches of human growth hormone (medical technology); Visceral gout among vultures (widespread use of diclofenac)
Microbial adaptation and change	Microbial evolution, response to selection in environment	Antibiotic-resistant bacteria, "antigenic drift" in influenza virus
Public health service failures	Curtailment or reduction in prevention programs; inadequate sanitation and vector control measures; failure of laboratories to contain pathogens	Resurgence of tuberculosis in the United States; cholera in refugee camps in Africa; resurgence of diphtheria in the former Soviet Union; Marburg Hemorrhagic Fever in 1967

Newly Emergent Diseases, Transmission Thresholds, and Globalization

In order to precipitate a deadly global epidemic, three disease facets must occur: high transmissibility with a relatively high basic reproduction number or R_0 (the R_0 is the number of secondary infections caused by an initial infection), high pathogenicity and lethality, and long-term environmental persistence or frequent reinfection from nonhuman host reservoirs to humans. The 2009 swine flu pandemic exhibited high transmissibility but lacked high pathogenicity and lethality. The 2014 West African outbreak of Ebola displayed relatively low transmissibility, yet very high pathogenicity and lethality. Both previous and recent epidemics suggest significant environmental persistence—in the 2014 case, within fruit bats. In contrast, the first and second plague pandemics, seemingly caused by pneumonic plague, exhibited high transmissibility and high pathogenicity and lethality. However, the third plague pandemic, which is bubonic plague, is typified by low high transmissibility, relatively low pathogenicity and lethality in the last 50 years, and yet persistent long-term environmental survival in various mammalian reservoirs.

4

Human-environmental relationships have proceeded through four great periods, and each has seen the emergence of new diseases (see Chapter 3, Local to Global impacts; McMichael 2004), and each has borne witness to the evolving nature of human-environmental thresholds and connectivity between fellow humans through trade and transportation that control disease transmission. The spatial extent and lethality of the three plague pandemics—the first pandemic of 541–715 CE, the second pandemic of 1330s–1879 CE, and the third and current pandemic that began in 1894 CE—illustrate these thresholds and changing connectivity. For instance, the first pandemic was restricted to Europe, the second to Europe and Asia, and the third occurs throughout the world.

The first great period of disease emergence occurred 5000–10,000 years ago when humans first became sedentary, living in permanent settlements, cultivating the land and domesticating animals. During this period, measles and pertussis first appeared among humans, suggesting a domesticated-animal origin; however, Pearce-Duvet argues that human modification of the environment might explain other crossovers such as tuberculosis. The second great period occurred 1000–3000 years ago when we began to trade across continents, for instance, the Silk Route and the Greek traders who followed Alexander the Great into India. The Silk Route repeatedly acted as conduit for the medieval Black Death between 1346 and 1815 as plague mutations erupted in Central Asia and were transmitted westward, infecting Europeans along the route a decade or so after their emergence. A third period began in 1492 as humans became intercontinental travelers, and in this case European settlers brought Native Americans measles, flu, and smallpox. For decades, it was thought that Native Americans gave Europeans syphilis, but four skeletons from fourteenth century in Hull, England, and evidence from victims of Vesuvius found in Pompeii suggest syphilis was already well established in Europe before Columbus set sail. Today, we live in the fourth great period of disease emergence, as our global transportation networks move goods and people and diseases, for example, SARS, rapidly around the globe. We live in an increasingly smaller world where millions of people move each day round the globe with comparative ease and frequency and could act, at any time, as perfect vectors for highly lethal, transmissible, newly emergent microorganism.

▶ **Example**

On January 12, 2010, a magnitude 7 earthquake struck Haiti, devastating the country and its people, killing an estimated 316,000 people and leaving nearly one million homeless. On October 21, 2010, cholera was identified in-country and quickly grew into an epidemic in a country that, in 2008, offered only 63% of its citizens improved drinking water and only 17% of its citizens access to improved sanitation. By December 2011, an estimated 522,335 cases of cholera and 7001 deaths were attributable to cholera. Similarly, on October 31, 2010, an outbreak of the same strain of *Vibrio cholerae* began in the Dominican Republic, where 86% of the population has access to improved drinking water and 83% to improved sanitation. In contrast, the Dominican Republic suffered 363 fatalities and 21,432 cases were reported. In 2010, 15 cases of *V. cholerae* serogroup 01, serotype Ogawa, biotype El Tor associated with the Hispaniola outbreak, were reported in the United States, and in 2011, an additional 40 cases were reported. In 2010 and 2011, 17 cases were identified in Florida, 13 in New York, 4 in Massachusetts, 3 each in Ohio and Texas, 2 each in Georgia, Kentucky, and Virginia, and one case in the following states of Arkansas, California, Illinois, Kansas, Missouri, New Jersey, New Mexico, and Pennsylvania. Most of these US cholera cases were tied to US citizens with Hispaniola roots returning to the USA after traveling to Hispaniola to provide technical, logistical, and food aid and emotional support in the aftermath of the earthquake. ◀

HIV-1

The global spread of HIV-1, AIDS, illustrates how spatial thresholds were exceeded within the fourth great transition. We suspect in precolonial Africa, HIV-1 infection rates were less than 1% across Central Africa where simian immunodeficiency virus (SIV) is found within the chimpanzee and gorilla populations in Cameroon, Democratic Republic of the Congo, Republic of the Congo, and Gabon. In fact, the most widespread HIV-1 clad is derived from a SIV_{cpz} found within chimpanzees, specifically those chimps found in SE Cameroon where the chimp SIV_{cpz} is identical to the oldest variation or clade of HIV-1 known to science.

During precolonial times, single sex partners were the norm; migration for work was very limited; transportation within, among, and between regions and countries extremely limited; and deforestation rates were very low. However, colonialism was a game-changer! Deforestation accelerated as colonial programs and multinationals encouraged poor to colonize peripheral forests and cut forest for commercial timber. Unfortunately, many monkeys and apes can still thrive in degraded forest and around human settlements. An unintended consequence we think is that the consumption of bushmeat increased and the transmission of monkey and ape viruses to humans increased. Colonial countries also increased the density of transportation links within countries and forced colonized peoples into a wage-labor market through taxation. Taxation forced colonized people to sell their time and labor for monetary reward. As a result, large numbers of African males in the early 1900s migrated to work-rich environments such as mines, ports, capitals, and other colonial towns. This also precipitated an increased in urban prostitution, facilitating an explosion in sexually transmitted diseases.

David Quammen postulates the following scenario: in the early 1900s a single human hunter is bitten or cut while killing a snared chimpanzee infected with SIV_{cpz}. This probably occurred countless times to countless hunters, but their infection did not cross to other individuals. With the advent of colonialism, mandatory taxation, and increased river trade and traffic, infected individuals had a greater likelihood of transmitting their acquired HIV-1. So, although cut and infected with SIV_{cpz}, the hunter, unaware of the infection, travels downstream into the DRC and infects one or more women in one or more of the river-trading towns in the upper Sangha River basin. These infected women unknowingly pass this SIV_{cpz}/HIV-1 onto subsequent male partners, who migrate further downstream. This movement was probably haphazard and intermittent and contingent on the long incubation period and infectiveness of HIV-1, and so, according to Worobey and coauthors, it was not before 1959/60 that individuals were infected in Kinshasa.

Thereafter, several additional factors assisted in the transmission of HIV-1 to a broader community. Needles and vaccines were first introduced into Africa in the post-WWII environment; however, resupply of needles and syringes was poor, sterilization procedures poor or nonexistent, and, as a result, needles and syringes were reused many hundreds, if not thousands, of times. Furthermore, in the postcolonial environment, political, ethnic, and economic rivals, either disenfranchised by the departing colonial administration or elevated above all others or false-fed propaganda by Cold War rivals, led to the advent of many civil conflicts and low-intensity warfare in the countryside. Poorly led and poorly paid soldiers frequently used beating, murder, and rape to intimidate rural people. Many rural peoples were dispossessed of land or land ruined by warfare, with the result that many of these people either migrated to towns or became involved in smuggling or prostitution. As a result, conditions were ripe for the accelerated transition of sexually transmitted diseases as multiple sexual partners became common place, famine and malnutrition stressed and weakened immune systems, large-scale migrations of people from rural to urban areas increased, and these migrations and low-intensity warfare and population growth devastated or overwhelmed healthcare systems across Central Africa.

By 1968, HIV-1 had crossed into the USA, quite possibly through Haiti and the blood-product trade for hemophiliacs, with the first known victim being a male prostitute in St. Louis. However, today in the USA, according to Adimora and Schoenbach, HIV-1 remains spatially concentrated in socially marginalized groups exhibiting low social capital, high incarceration rates, and high-income inequality.

In contrast to influenza and the cold virus, HIV-1 is difficult to transmit—it is transmitted blood-to-blood rather than through viral aerosols issuing from the lungs through coughing. We now know that there was not a patient zero identified by the world's vitriolic media as Gaetan Dugas, a Canadian flight attendant; rather,

HIV-1 transmission velocities were quite slow. In contrast, the 1918 influenza epidemic engulfed the world in two years and killed more victims than WWI—this at a time of a relatively slow oceanic traffic in the form of ocean streamers. Today, we are no more than 48 hours from any place on earth. This is clearly illustrated by the recent global eruptions of SARS, swine flu, and now COVID-19. Whereas SARS was ultimately contained, although three super-spreader centers facilitated transmission beyond Hong Kong, COVID-19 escaped Wuhan, China, and as we write in March 2020, it is still expanding as a global pandemic with ~1 million infections and ~48,000 deaths. What crucially facilitated the emergence of the ongoing COVID-19 pandemic is that Wuhan, China, is fully embedded within the global transportation network while it was not in 2003 during the SARS epidemic. Wuhan has an international airport with direct flights to Europe and much of Asia. Wuhan is also part of China's growing high-speed rail network. In other words, China crossed a transportation threshold sometime between 2003 and today making the likelihood of newly emergent diseases escaping the confines of China all the more likely.

As illustrated by 1918 flu, swine flu, seasonal flu, SARS, and now COVID-19, we long ago as a global community crossed a social distance threshold, whereby pneumonic diseases that spread by aerosol sprays of viruses and bacteria through coughs and sneezes and the subsequent short-term persistence of these pathogens on interior surfaces of building (and our propensity to touch faces) can be transmitted around the world with ease. Our cities are *sadly* perfect breeding grounds for new emergent pneumonic pathogens! Urbanites live in high densities and endure nearly constant social interactions. Our proclivity to work in crowded office spaces, teach face-to-face, commute on crowded trains and buses, eat and drink out, and work-out in frequently crowded gyms places urbanites at greater risk than those individuals living in rural areas. The critical issue with social distancing according to Rabinowitz and Bartman is that the more viral particles a person is exposed to, the greater the likelihood that the pathogen will overcome an individual's immune system defenses and infect an individual. Simply put, the more social contacts a person is subjected to during the course of a day, week, month, or year, the greater the likelihood that the individual will become infected if the pathogen resides within the confines of that individual's social space. This explains why even nurses and doctors who take precautions while treating infected individuals still have a high likelihood of becoming infected themselves. This also explains why working-class individuals living in high-density housing also suffer higher disease risks—we are seeing both these phenomena in New York City.

Climate, Disease, and Famine

New and well-dated evidence of sulfate deposits in Greenland and Antarctic ice cores indicates a substantial and extensive atmospheric acidic dust veil at AD 533–534 ± 2 years. This was likely produced by a large explosive, near equatorial volcanic eruption, causing widespread dimming and contributing to the abrupt cooling across much of the Northern Hemisphere. Two additional eruptions in 540 and 547 CE transformed the Late Antique World by initiating the Late Antique Little

Ice Age that ran from 536 to 660 CE. Tree-ring data suggest that this was the most severe and protracted short-term cold episode across the Northern Hemisphere in the last 2000 years. Cold and subsequent droughts led to Mediterranean, Mesopotamian, and Chinese famines, and, according to P. D. Pang and associates, deaths up to 70–80% of the Chinese population (Pang et al. 1989). Rather, opportunistically, the Justinianic Plague—*Yeresina pestis*—erupted among Eurasia's impoverished, malnourished, and health-compromised peoples. Perhaps as many as half the population of the Byzantine Empire and Europe was killed between 542 and 565 CE.

Support for this contention comes from the Nazi-instigated Dutch famine of 1944–1945. This famine compromised Dutch health for decades, especially among those children born or conceived immediately before or during this famine. McMichael and coauthors found that climate anomalies directly impact health through heat waves, dust, or extreme weather, or indirectly by reducing agricultural yields, reducing water supply, triggering eruptions of climatic-sensitive infectious diseases, displacing human groups, and triggering depression and despair among marginalized, disadvantaged, and failing farm communities.

Given the impoverished, malnourished condition and recently climate-displaced nature of many peoples around the world, are we approaching a critical disease/famine threshold? Certainly, the recent cholera epidemic in Haiti triggered by the chaos of the post-earthquake environment suggests we are ill-equipped to deal with any and all widespread societal problems that climate change is initiating. The recent and continuing collapse in insect populations around the world is equally troubling—insect pollination of the world's agricultural plant crops is at considerable risk. Although it is unlikely we have a reached a critical insect density pollination threshold, we are certainly moving toward a critical threshold triggered by climate change, the widespread use of neonicotinoids insecticides, and seemingly uncontrolled sprawl.

4.5 Can We Prevent a Future Pandemic?

Enacting civilian air flight bans across the USA has been proposed as a means to limit the spread of pandemic diseases such as influenza. The decline in air traffic post 9/11 appeared to explain 27% of the lag in peak of the influenza epidemic that winter/spring (Brownstein et al. 2006). A more extensive meta-analysis found no difference between the spread of the post 9/11 influenza epidemic than influenza epidemics from 1972–2002 (Viboud et al. 2006).

Pandemic simulations suggest that isolating just 2% of the largest global cities dramatically reduce the transmission of infectious diseases. However, 27.5% of the principal airline network nodes (e.g., Atlanta International, Heathrow, Dallas-Fort Worth, Frankfurt) had to be shut down to obtain a similar response (Hufnagel et al. 2004). The recent 2003–2004 SARS epidemic illustrated that a few infected individuals can infect a high number of susceptibles; these super-spreading events (SSEs) can also rapidly transmit infectious diseases across very heterogeneous spaces. In Hong Kong, three SSEs were identified; one in a hospital, one in a housing estate, and one at a hotel. Other SARS-related SSEs occurred in Beijing, Singapore, and Canada. There is also a strong, positive relationship between relative population concentrations of residents with self-identified ties to Hispaniola and US cholera cases during 2010–2011. This suggests where strong social, cultural, and economic connections exist between places, highly infectious diseases might easily jump across spatial boundaries and move up spatial scales.

As is becoming increasingly transparent, the global community is in the process of failing to prevent a Coronavirus-19 or COVID-19 pandemic. Yet the first cluster of unusual pneumonia cases in Wuhan was identified by an Artificial Intelligence platform BlueDot a little after midnight on Dec 30, 2019 (Niiler 2020)—this some six days before the CDC sent out an alert, and some nine-days before the World Health Organization alerted the world to a novel, newly emergent coronavirus. Yet businesspeople, tourists, and travelers of all sorts continued to fly in and out of China, South Korea, and Italy, three hotspots in the COVID-19 epidemic until mid-March 2020. Most US educational institutions have recalled students on study abroad trips from these destinations. Italy attempted to seal off Lombardy, and initially this worked, but evidence suggests this lock-down was leaky at best! But on March 8, quarantined Lombardy, states around Venice, and the Rimini area effectively sealed in 16 million, shutting down all schools and universities. The long incubation period of COVID-19 has certainly impeded efforts to quarantine areas. As of March 6, 2020, it is thought that COVID-19 was circulating within Washington State, USA, for a few weeks before being detected at a nursing home for the elderly.

What swine flu of 2009 (H1N1), SARS, Middle East Respiratory Syndrome (MERS), and COVID-19 illustrate is that, today, there are few barriers to a pandemic if a human-to-human pathogen is highly transmissible, in other words, (1) exhibits a high level of infectivity, (2) has an incubation period of 5–7 days or more, (3) causes asymptomatic infectious carriers, in other words, people who appear healthy, remain mobile, but infect others, and (4) appears to exhibit a low level of lethality. So, although the world's media is fixated on the coronavirus (as of March 6, 2020) and scientists, politicians, and laypeople are genuinely and rightfully worried about coronavirus, unless very strict quarantines had been imposed immediately upon its discovery, our full globalized society (where every place on earth is 48 hours away from every other place) has crossed a ***pandemic threshold***. In contrast, where pathogens exhibit high lethality and short incubation periods (victims are easily identified and too ill to travel any distance), public health systems can establish effective quarantines. But even under these circumstances, cultural, political, and economic conditions can sometimes optimize local or regional epidemics (rather than a global pandemic) such as the Ebola epidemic that killed over 11,000 in and around Sierra Leone in 2014.

Even by early March 2020, the COVID-19 pandemic had slowed economic activity in China, reduced pollution above China, and triggered a global economic crisis. According to Kasha Patel and NASA, February's NO2 levels above China were 10–30% lower than just a month before.

4.6 Global Climate Thresholds and Tipping Points

According to many climate researchers, we are approaching or at the last exit ramp where we can make significant changes to our lifestyles and economic systems that could avert runaway global warming. Yet within the last century, atmospheric temperatures have risen 2.5 °C and the oceans pH has dropped 0.1 units as they have absorbed CO2; as a result, in excess of 35% of the world's coral reefs have bleached and died due to heat-stress and acidification. Estimates suggest 90% of coral reefs will die by 2050. This will have a crippling effect on tropical fish fecundity as many young tropical fish rely on living coral reefs for protection.

Ignoring global climate change, first hypothesized in the late nineteenth century but first measured scientifically in the late 1930s and widely acknowledged after 1960 with the publication of Keeling Curve, appears now to have been extremely short-sighted; Mayer Hillman goes further and suggests we are doomed:

> The outcome is death, and it's the end of most life on the planet because we're so dependent on the burning of fossil fuels. There are no means of reversing the process which is melting the polar ice caps. And very few appear to be prepared to say so. Barkham, 26 April 2018.

The controversial identification of the "hockey-stick" acceleration in CO_2 and associated temperatures for the late twentieth century by Mann and colleagues in 1998 illustrates how societies, even our modern global society, can conflate, repudiate, attack, and dismiss documented evidence that they are in deep trouble: among the chief climate change deniers are a very vocal array of US citizens, the Republican Party, and Fox News. Yet we have talked to rural, isolated peoples in Ecuador, India, and PNG who have witnessed climate change during their own lifetime and acknowledge its global dimension (see ◻ Fig. 4.1). Since 1998, broad scientific consensus has identified unprecedented increases in CO_2 and associated global warming. For instance, the UK Met Office notes that 16 of the 17 warmest years on record have occurred since 2000.

Tipping Points, Irreversible Changes, and the IPCC

The IPCC's Third and Fourth Assessment Reports introduced the notion of tipping points within the earth-atmosphere system whereby abrupt and/or irreversible change will occur if anthropogenic warming exceeds 5 °C above pre-industrial levels. This estimate has proved to be too conservative, and encouraged industries, laypeople, and politicians across the world to be rather tentative in their approach to reducing GHG emissions. Recent IPCC reports *Global Warming of 1.5 °C* and *IPCC Special Report on the Ocean and Cryosphere in a Changing Climate* dispel this notion and suggest that abrupt and irreversible change will occur if anthropogenic warming exceeds just 1 °C warming. Yet this is where we reside today! Furthermore, these changes will be nonlinear, unpredictable, and far-reaching, exhibiting teleconnections across the planet (just as El Niño does today).

Climate change is expected to be increasingly nonlinear once a number of thresholds or tipping points are exceeded. Arctic sea ice loss is weakening the

Arctic jet, allowing more poleward penetration of warm continental air masses into the Arctic but also triggering more intense polar vortexes. Both the warming Arctic and melting Greenland ice sheet are slowing down the meridional overturning circulation (MOC) due to the influx of surficial freshwater into the North Atlantic. To clarify, the MOC stores and transports heat, freshwater, and carbon around the globe. The slowing down of the MOC is reducing CO2 storage in the deep ocean. In fact, Bertler and coauthors suggest that between 1550 CE and 1800 CE, atmospheric CO_2 decreased by 10 ppmv as more CO2 was absorbed by the deep MOC adjacent to Antarctica. Today, Greenland is losing ice at the fastest rate since the end of the Little Ice Age 350 years ago. In the North Atlantic, just off the coast of Greenland, the cooling MOC loses buoyancy as heat is lost from the surficial water (this surficial water originates as glacial meltwater from the Greenland Ice Cap) to the atmosphere. The MOC current then flows on the deep ocean floor towards the Southern Atlantic, where upwelling returns the nutrient-rich MOC to the ocean's surface adjacent to Southern Africa. A slower MOC stores less carbon and hence fewer nutrients, less heat storage, and less freshwater. As a result, the Arctic is becoming warmer and drier, causing more droughts in the boreal forest area of NA. A similar scenario plays out across the Amazon basin, although this is being intensified by a recent acceleration in deforestation as more forest is being cut and burnt to provide immediate acreage for organic soybean cultivation. The slowing MOC is reducing upwelling of cold water off both Antarctica and Southern Africa, which is intensifying Antarctic ice sheet melting and reducing nutrient-rich sources along the edge of both continents, with catastrophic impacts on commercial fishing, and krill and sardine numbers that are impacting whales, penguins, albatrosses, and seals, among others, that thrive in these formally carbon-rich waters.

Just to confuse things, massive melting of the Greenland ice sheet and subsequent intrusion in the North Atlantic of buoyant, freshwater water have in the past been associated with the rapid onset on the Mini Ice Age following the peak of the Medieval Warm Period. The hypothesis was that buoyant, low-salinity freshwater flowing off Greenland forces the surficial North Atlantic drift south, denying Europe warm water and warm oceanic air masses. The rapid cooling of Europe and subsequent droughts and famines created conditions that compromised the health and immune systems of the majority of the European population and thus facilitated the transmission and lethality of plague or Medieval Black Death once it arrived in Europe, courtesy of the Golden Horde's use of plague victims as biological weapons in the siege of Caffa in 1346.

Lenton and coauthors argue that the collapse of both the Amundsen Sea embayment of West Antarctica and the Wilkes Basin of the East Antarctic ice sheet are imminent! Combined, they could trigger a 6–7 meter rise in sea level somewhere within the next 100–1000 years. However, the Greenland ice sheet is doomed and expected to collapse by 2030. Furthermore, the Arctic is expected to become ice-free in summer with 2 °C warming. Elsewhere, a 99% die-off of corals is expected if global temperatures reach 2 °C above pre-industrial levels. The impact on all marine commercial fisheries would be catastrophic, while the ocean's principle tropical marine fish hatchery would be lost. Lenton and coauthors also suggest that the Amazon's tipping point, where negative atmospheric feedbacks would shrink moisture inputs triggering desertification, lies somewhere between 20–40% forest-cover loss. Since 1970, 17% of the Amazon forest has been lost, yet deforestation is currently accelerating after years of decline. The rapid warming of the Arctic has led to pest infestations, fires, and an increase in the number and intensity of boreal forest droughts.

4

Estimates vary, but Lenton and coauthors suggest that in order to remain below 1.5 °C warming, we must cap CO2 emissions and not emit another 500 gigatons (Gt) of CO2. Sadly, we are already close to this threshold! Permafrost emissions are expected to yield 100 Gt of CO2 and unquantified amounts of methane. While the Amazon is expected to yield 90 Gt of CO2 and the boreal forests another 110 Gt of CO2. However, we are currently emitting 36–40 Gt of CO2 annually, suggesting we will exceed the 500 Gt limit in the 5–7 years!

4.7 Climate, Tipping Points, and Mass Mortality Events

Although extinctions will be addressed in detail in Chap. 8, localized extinctions or mass mortality events (MMEs) suggest that many regions on earth have reached critical ecological thresholds that could lead to broader-scale global extinctions. An ecological threshold is the condition where a relatively small change or disturbance in the landscape, for instance, an increase in the frequency and duration of droughts, can rapidly change the functioning and structure of an ecosystem. Both Scotland and Alaska were once world famous for their salmon fishing; in January 2017, the Alaskan salmon fishery collapsed, and in 2018, the Scottish salmon fishery also collapsed! For instance, not a single salmon was caught within the Sprey and Nith rivers in 2018. Katherine Mills and coauthors suggest these declines started in the 1960s, but unlike the decline in trout observed in Norway which appears connected to increased lake acidity caused by sulfur emissions from coal-burning in the UK power-plants, salmon collapses appear linked to warmer ocean temperatures that are negatively impacting plankton communities.

Declining plankton communities and harmful algal blooms (HABs) appear to have recently triggered mass mortality events among seabirds in the North Pacific and North Atlantic. For instance, between October 2016 and January 2017, 350 dead puffins and other seabirds washed up on the shore of Saint Paul Island. In fact, Piatt and coauthors suggest some 62,000 dead or dying common murres (*Uria aalge*) washed up on beaches from California to Alaska between the summer of 2015 and the spring 2016, of which two-thirds were adults. In reality, possibly as many as a million birds died because very few (ocean-wandering) dead or dying birds will wash ashore. Subsequently, 22 murre breeding colonies completely failed between 2015 and 2017. This is unprecedented!

Timothy Jones and coauthors suggest that a period of elevated sea-surface temperatures in the eastern Bering Sea triggered a mass mortality event in surficial zooplankton and a reduction in prey-fish densities that precipitated a mass mortality event or mass die-off among fish, bird, and mammal species in 2014–2017. Piatt and coauthors nicknamed this record-breaking, super-heated pool of surface water in the eastern Bering Sea, a "Blob." The Blob was a product of surficial heating that began in 2013, continued through the 2015 El Nino, and lasted through to 2016—all a product of global warming. The Blob stretched along the NW US coastline for 1600 km and exhibited anomalous sea-surface temperatures 3–6 °C above normal. According to Jessie Yeung, in September 2019, another Blob was discovered off Washington State and yet another Blob east of New Zealand some 1 million square km in size and exhibiting SST 5 °C warmer than normal.

It appears that mass mortality events triggered by climate events are on the rise and are not restricted to fish or birds! Mass die-offs of saiga antelope in Kazakhstan occurred in 1981 with 70,000 dead, again in 1988 with 200,000 dead, and again in 2015 with 200,000 dead! Postmortem analysis in 2015 concluded that *Pasteurella multocida,* that harmlessly infects most antelopes, erupted to kill thousands. Recent work by Kock and coauthors suggests that these mass die-offs occurred following heat waves of greater than 37 °C, with in excess of 80% humidity suggesting climate triggered a threshold event among the bacteria allowing them to become vascular and lethal.

In 2019, over 100 gray whales washed up dead along the western seaboard of the USA. This represents the largest mass die-off of gray whales recorded. Many that were found were emaciated! In one case, a whale had resorted to eating eel grass, and yet eel grass offers nothing nutritionally to whales. Several dead whales have been found hundreds of miles from traditional feeding areas and migration routes suggesting the whales were desperately searching for food. Together, these facts suggest climate change and warming sea temperatures are changing the location and supply of whale prey creatures such as amphipods, small, bottom-dwelling oceanic shrimps.

Puffins, saiga antelope, and gray whales mass die-offs can be tied to some form of environmental or biotic perturbation triggered by climate change. Other mass die-offs occur over longer time periods and hence are more difficult to establish a cause-and-effect. Nevertheless, Christmas Bird Counts, Birding Bird Surveys, Big Day Counts, Backyard Counts sponsored by the likes of the British Trust for Ornithology and the RSPB in the UK, and state DNRs, Audubon Society, and the Cornell Lab for Ornithology in the USA, suggest ramped, slow climate change is responsible for population crashes among house sparrows in the UK and evening grosbeaks in the North America. Interestingly, the house sparrow in NA appears to be expanding, yet its mother-population in Europe is struggling. The RSPB offers hints and plans for house sparrow nest boxes to its membership to help reduce their mortalities.

It is possible that the 2019/2020 bushfires in Australia have reached a tipping point. Barker and Price show that within eucalypt forests, crown fires are twice as likely after a previous crown fire than an understory fire. Similarly, understory fires are more frequent if preceded by an understory fire rather than a crown fire. These relationships mirror fires observed in NA, suggesting there are fire thresholds! According to Bowman and coauthors, the mature alpine ash forest of Australia is critically threatened by fires. Since 2002, approximately 85% of the alpine forest has been burnt, and in many areas suffered repeat burns. These repeat burns have killed 97% of all alpine ash seedlings. Analysis, in 2009 by Leonard and coauthors, of the Kilmore-Murrindindi fire complex that burnt ~250,000 ha in Victoria, Australia, suggests that the number and size of unburnt patches of forest, that act as floral and faunal forest refuges, are determined by topographic and vegetational micro-climates yet fuel load and broader climate variables can override these effects. Their work suggests that the excessive heating that Australia suffered before and during the 2019/2020 fire season will reduce the number of surviving refuges. As a consequence, across Australia, a divergent floral landscape is beginning to replace the mature forests.

In contrast, today some 3200 pairs of Dartford warbler establish territories, which barely bred on the UK mainland in the 1960s, with only 22 pairs breeding in 1966. Gibbons and Wotton suggest a long-run of mild winters since 1963, the product of global warming, lowered the climate-driven mortality threshold of Dartford warblers that exist at the outer limit of their range in the UK.

4.8 Deforestation Tipping Points

According to Nowosad and Stepinski, approximately 15% of the earth's surface between 1992 and 2015 was modified through either direct human action or climate-induced changes, much of this from natural forest to agriculture. But much of this occurred as forest fragments left after previous episodes of deforestation (identified at the scale of 9 km × 9 km or 81 km^2 blocks) were converted to agricultural land-use. Their work suggests recent deforestation activities mirror deforestation practices observed across the American Midwest in 1830–1920s, southwestern Australia in the 1950s, and the UK in 1914–1945, which left less than 5% of England forested. Nowosad and Stepinski found globally that deforestation rapidly accelerated once when more than 50% of the land surface of each 81 km^2 block was deforested. Across the tropics, deforestation driven by the forest frontier phenomenon (see Climate chapter) where inverse urban-rural migration seeded by tropical forest transition policies of the 1960s–1970s in South America and income inequality (see Extinction chapter) drive the slash-and-burning of tropical forest adjacent to newly built roads, then in-fill between roads, and finally precipitate further road-extension into the forest as the *land gives out*. As Miller notes, once penetration into the forest occurs, subsequent forest degradation makes it comparatively easy to finish the job of forest elimination.

It seems humans despise heterogeneous landscapes of farms and scattered forest fragments!

Lovejoy and Nobre support these contentions and suggest that the surviving Amazon rainforest has reached a critical tipping point that once breached will initiate an irreversible *dieback* process, whereby less and less rainfall will fall on the Amazon. At the moment, 75% of all rain falling on the Amazon is returned to the atmosphere above the forest through transpiration—the Amazon is hot and humid! Incredibly, the forest recycles this moisture five to six times before air masses move this moisture up the Andes. As more of the Amazon is deforested, less moisture will be recycled. According to Lovejoy and Nobre, currently 17% of the entire Amazon basin has been deforested and rainfall across the Amazon is already declining. As deforestation continues, this will lead to the development of a savannah-like landscape of scattered trees and huge expanses of grass replacing the once giant tropical rainforest. The impact of carbon sequestering will be catastrophic. This will accelerate climate change. But this is not the only issue! The drying and opening of the Amazon rainforest and population by peoples has significantly increased the fires that erupt during its moderate dry season. Although the 2019 fire season was not the worst—2005 was two-and-half times bigger in spatial extent and number of fires—Global Fire Database reports that the Amazon is burning at rates far in excess of fires observed in the 1960s, 1970s, and 1980s.

It appears that the Amazon and Congo are about to cross a catastrophic tipping point; within the next decade, according to Hubau and coauthors, both the Amazon and Congo rainforests will no longer act as a "carbon sinks." If this happens, this will greatly accelerate global climate change. Previous work suggested that the Congo was a stable, net carbon sink, but their work suggests that the Congo and other tropical African forests are in decline. They also suggest that forest uptake of carbon peaked in the 1990s and has been in steady decline since.

4.9 Soils and Crop Production Thresholds

Today, vast amounts of chemical fertilizer are necessary to keep crop production levels at or near levels attained 20–30 years ago even in Iowa, Cambridgeshire, the Netherlands, Ukraine, paddy fields of India, or Mexico. However, with rigorous management, organic farming can attain and sometimes exceed 80% production levels of non-organic commercial farming.

▶ Example

Joel Simon writing on the crisis in El Campo graphically illustrated the impact of political land reform and use of fertilizer as bribes in Mexico on rural-urban, rural-international migrations, and drug cartels. Before the Spanish conquest, the mountains of Oaxaca were heavily forested and sustained a corn culture and economy. Today, the earth no longer gives! The land cannot support corn nor peoples. The mountains of Oaxaca are among the most heavily eroded landscapes on earth, 70% of the state's arable land is ruined, and many of its peoples have fled into the barrios of Mexico City, or moved on to the USA, or been absorbed into Mexico's deadly drug war.

And yet Mexico's Green Revolution that began in the 1950s promised such a different outcome. Simon found that the Green Revolution reached the lush hills and mountains of Oaxaca in the mid-1970s. Government officials donated several bags of chemical fertilizer each year to each farmer, in return for their loyalty at each election. Initially, the corn stalks grew larger and more quickly, but corn crops failed if chemical fertilizer was not applied each year. Simply put, the additional artificial fertilizer stripped the remaining natural fertility of the soils away as the corn grew bigger and faster. At this point, the farmers and their corn culture were tied inextricably to chemical fertilizer! The soil's natural fertility threshold had been exceeded. In fact, more fertilizer was needed each year to maintain corn yields. Yet, critically, the free program was canceled and instead replaced by Banrural, a government farm bank that offered credit lines for fertilizer. But each year, more fertilizer was needed, so by the late 1990s, according to Simon, 80–90% of the working-aged males in communities like San Juan Mixtepec in Oaxaca were working in the USA to pay farm debts incurred in trying to maintain their traditional corn culture. Today, labor migration from Mexico to the USA is difficult and sometimes deadly, and so more Oaxacans are left mired in debt in Mexico. ◀

References

Adimora, A. A., Schoenbach, V. J., & Doherty, I. A. (2006). HIV and African Americans in the southern United States: Sexual networks and social context. *Sexually Transmitted Diseases, 33*(7), S39–S45.

Alroy, J. (2001). A multispecies overkill simulation of the end-Pleistocene megafaunal mass extinction. *Science, 292*(5523), 1893–1896.

Asquith, N. M., Wright, S. J., & Clauss, M. J. (1997). Does mammal community composition control recruitment in neotropical forests? Evidence from Panama. *Ecology, 78*(3), 941–946.

Barker, J. W., & Price, O. F. (2018). Positive severity feedback between consecutive fires in dry eucalypt forests of southern Australia. *Ecosphere, 9*(3), e02110.

Barkham, P. (2018). 'We're doomed': Mayer Hilland on the climate reality no one else will dare mention. The Guardian, 26 April 2018. Accessed 24 Aug 2018.

Bertler, N. A. N., Mayewski, P. A., & Carter, L. (2011). Cold conditions in Antarctica during the little ice age—Implications for abrupt climate change mechanisms. *Earth and Planetary Science Letters, 308*(1), 41–51.

Bowman, D. M., Murphy, B. P., Neyland, D. L., Williamson, G. J., & Prior, L. D. (2014). Abrupt fire regime change may cause landscape-wide loss of mature obligate seeder forests. *Global Change Biology, 20*(3), 1008–1015.

Brownstein, J. S., Wolfe, C. J., & Mandl, K. D. (2006). Empirical evidence for the effect of airline travel on inter-regional influenza spread in the United States. *PLoS Medicine, 3*, e401. https://doi.org/10.1371/journal.pmed.0030401.

Brunsden, D., & Thornes, J. B. (1979). Landscape sensitivity and change. *Transactions of the Institute of British Geographers, 4*(4), 463–484.

Callendar, G. S. (1938). The artificial production of carbon dioxide and its influence on temperature. *Quarterly Journal of the Royal Meteorological Society, 64*(275), 223–240.

Diamond, J. (2005). *Collapse: How societies choose to fail or succeed*. New York: Penguin.

Ehrlich, P. R. (1988). The loss of diversity. *Biodiversity*. In *National Academy of Sciences* (pp. 21–22). Washington, DC: National Academies Press.

Findlay, C. S., & Bourdages, J. (2000). Response time of wetland biodiversity to road construction on adjacent lands. *Conservation Biology, 14*(1), 86–94.

Firestone, R. B., West, A., Kennett, J. P., Becker, L., Bunch, T. E., Revay, Z. S., Schultz, P. H., Belgya, T., Kennett, D. J., Erlandson, J. M., & Dickenson, O. J. (2007). Evidence for an extraterrestrial impact 12,900 years ago that contributed to the megafaunal extinctions and the younger Dryas cooling. *Proceedings of the National Academy of Sciences, 104*(41), 16016–16021.

Forman, R. T., & Deblinger, R. D. (2000). The ecological road-effect zone of a Massachusetts (USA) suburban highway. *Conservation Biology, 14*(1), 36–46.

Gibbons, D. W., & Wotton, S. (1996). The Dartford warbler in the United Kingdom in 1994. *British Birds, 89*(5), 203–212.

Goudie, A. (2006). Global warming and fluvial geomorphology. *Geomorphology, 79*, 384–394.

Harvey, A. M. (1992). Process interactions, temporal scales and the development of hillslope gully systems: Howgill fells, Northwest England. *Geomorphology, 5*(3–5), 323–344.

https://www.theguardian.com/environment/2018/apr/26/were-doomed-mayer-hillman-on-the-climate-reality-no-one-else-will-dare-mention?CMP=fb_gu

Hubau, W., Lewis, S. L., Phillips, O. L., Affum-Baffoe, K., Beeckman, H., Cuní-Sanchez, A., Daniels, A. K., Ewango, C. E. N., Fauset, S., Mukinzi, J. M., & Sheil, D. (2020). Asynchronous carbon sink saturation in African and Amazonian tropical forests. *Nature, 579*, 80–87. https://doi.org/10.1038/s41586-020-2035-0.

Hufnagel, L., Brockman, D., & Geisel, T. (2004). Forecast and control of epidemics in a globalized world. *PNAS, 101*(42), 15124–15129.

Hung, C. M., Shaner, P. J. L., Zink, R. M., Liu, W. C., Chu, T. C., Huang, W. S., & Li, S. H. (2014). Drastic population fluctuations explain the rapid extinction of the passenger pigeon. *Proceedings of the National Academy of Sciences, 111*(29), 10636–10641.

IPCC, 2018. Global warming of 1.5°C. (IPCC).

IPCC. (2019). *IPCC Special Report on the Ocean and Cryosphere in a Changing Climate (IPCC)*. Geneva: Intergovernmental Panel on Climate Change.

James, L. A. (1993). Sustained reworking of hydraulic mining sediment in California: GK Gilbert's sediment wave model reconsidered. *Zeitschrift für Geomorphologie, Supplementbände, 88*, 49–66.

Johnson, N. K. (1975). Controls of number of bird species on montane islands in the Great Basin. *Evolution, 29*(3), 545–567.

Jones, T., Divine, L. M., Renner, H., Knowles, S., Lefebvre, K. A., Burgess, H. K., Wright, C., & Parrish, J. K. (2019). Unusual mortality of tufted puffins (Fratercula cirrhata) in the eastern Bering Sea. *PLoS One, 14*(5), e0216532.

Karr, J. R. (1982). Population variability and extinction in the avifauna of a tropical land bridge island. *Ecology, 63*(6), 1975–1978.

Keeling, C. D. (1960). The concentration and isotopic abundances of carbon dioxide in the atmosphere. *Tellus, 12*(2), 200–203.

Knorr, W., Prentice, I. C., House, I. J., & Holland, E. A. (2005). On the available evidence for the temperature dependence of soil organic carbon. *Biogeosciences Discussions, 2*(4), 749–755.

Knox, J. C. (1993). Large increases in flood magnitudes in response to modest changes in climate. *Nature, 361*, 430–432.

Kock, R. A., Orynbayev, M., Robinson, S., Zuther, S., Singh, N. J., Beauvais, W., Morgan, E. R., Kerimbayev, A., Khomenko, S., Martineau, H. M., & Rystaeva, R. (2018). Saigas on the brink: Multidisciplinary analysis of the factors influencing mass mortality events. *Science Advances, 4*(1), eaao2314.

Lenton, T. M., Rockström, J., Gaffney, O., Rahmstorf, S., Richardson, K., Steffen, W., & Schellnhuber, H. J. (2019). Climate tipping points—Too risky to bet against. *Nature, 575*, 592–595.

Lovejoy, T. E., & Nobre, C. (2019). Amazon tipping point: Last chance for action. *Science Advances, 5*, eaba2949. https://doi.org/10.1126/sciadv.aba2949.

McLaughlin, S. P. (1994). An overview of the flora of the Sky Islands, southeastern Arizona: diversity, affinities, and insularity. *Biodiversity and management of the Madrean Archipelago: the Sky Islands of southwestern United States and northwestern Mexico*, pp.60–70.

McMichael, A. J., Woodruff, R. E., & Hales, S. (2006). Climate change and human health: Present and future risks. *The Lancet, 367*(9513), 859–869.

Met Office. (2017). 2016: One of the warmest two years on record. 18 January 2017. Accessed 27 Aug 2018. https://www.metoffice.gov.uk/news/releases/2017/2016-record-breaking-year-for-global-temperature

Mieth, A., & Bork, H. R. (2005). History, origin and extent of soil erosion on Easter Island (Rapa Nui). *Catena, 63*(2–3), 244–260.

Miller, M. 2020. Geographers find tipping point in deforestation. Phys Org News, Accessed 9 Jan 2020. https://phys.org/news/2020-01-geographers-deforestation.html

Mills, K. E., Pershing, A. J., Sheehan, T. F., & Mountain, D. (2013). Climate and ecosystem linkages explain widespread declines in North American Atlantic salmon populations. *Global Change Biology, 19*(10), 3046–3061.

Morse, S. S. (2001). Factors in the emergence of infectious diseases. In *Plagues and politics* (pp. 8–26). London: Palgrave Macmillan.

Niiler, E. (2020). An AL epidemiologist sent the first warning of the Wuhan virus. Wired, 25 Jan 2020. https://www.wired.com/story/ai-epidemiologist-wuhan-public-health-warnings/

Nowosad, J., & Stepinski, T. F. (2019). Stochastic, empirically-informed model of landscape dynamics and its application to deforestation scenarios. *Geophysical Research Letters*. https://doi.org/10.1029/2019GL085952.

Pang, K. D., Keston, R., Srivastava, S. K., & Chou, H. H. (1989). Climatic and hydrologic extremes in early Chinese history: Possible causes and dates. *Eos, 70*, 1095.

Patel, K. (2020). Airborne Nitrogen Dioxide plummets over China. https://earthobservatory.nasa.gov/images/146362/airborne-nitrogen-dioxide-plummets-over-china

Pearce-Duvet, J. M. (2006). The origin of human pathogens: Evaluating the role of agriculture and domestic animals in the evolution of human disease. *Biological Reviews, 81*(3), 369–382.

Phillipps, Q. (2016). *Phillipps' field guide to the mammals of Borneo and their ecology: Sabah, Sarawak, Brunei, and Kalimantan* (Vol. 105). Princeton University Press.

Phillips, J. D. (2009). Changes, perturbations, and responses in geomorphic systems. *Progress in Physical Geography, 33*(1), 17–30.

Phillips, J. D., & Van Dyke, C. (2016). Principles of geomorphic disturbance and recovery in response to storms. *Earth Surface Processes and Landforms, 41*(7), 971–979.

Piatt, J. F., Parrish, J. K., Renner, H. M., Schoen, S. K., Jone, T. T., Arimitsu, M. L., Kuletz, K. J., Bodenstein, B., García-Reyes, M., Duerr, R. S., Corcoran, R. M., Kaler, R. S. A., McChesney, G. J., Golightly, R. T., Coletti, H. A., Suryan, R. M., Burgess, H. K., Lindsay, J., Lindquist, K., Warzybok, P. M., Jahncke, J., Roletto, J., & Sydeman, W. J. (2020). Extreme mortality and reproductive failure of common murres resulting from the Northeast Pacific marine heatwave of 2014-2016. *PLoS One*. https://doi.org/10.1371/journal.pone.0226087.

Quamen, D. (1996). *The song of the dodo. Island Biogeography in an Age of Extinction* (p. 702). London: Pimlico.

Quammen, D. (2014). *The chimp and the river*. New York: Norton.

Sazonova, T. S., Romanovsky, V. E., Walsh, J. E., & Sergueev, D. O. (2004). Permafrost dynamics in the 20th and 21st centuries along the east Siberian transect. *Journal of Geophysical Research: Atmospheres, 109*, D1.

Schumm, S. A. (1973). Geomorphic thresholds and complex response of drainage systems. In M. Morisawa (Ed.), *Fluvial geomorphology* (pp. 299–310). Binghamton: State University of New York Publication of Geomorphology.

Simon, J. (1997). *Endangered Mexico: An environment on the edge*. Latin America Bureau: A Sierra Club Book.

Stott, T. A., & Mount, N. J. (2007). Alpine proglacial suspended sediment dynamics in warm and cool ablation seasons: Implications for global warming. *Journal of Hydrology, 332*, 259–270.

Syphard, A. D., Radeloff, V. C., Keeley, J. E., Hawbaker, T. J., Clayton, M. K., Stewart, S. I., & Hammer, R. B. (2007). Human influence on California fire regimes. *Ecological Applications, 17*(5), 1388–1402.

Viboud, C., Miller, M. A., Grenfell, B. T., Bjørnstad, O. N., & Simonsen, L. (2006). Air travel and the spread of influenza: Important caveats. *PLoS Medicine, 3*(11), e503. https://doi.org/10.1371/journal.pmed.0030503.

Voosen, P. (2018). Massive crater under Greenland's ice points to climate-altering impact in the time of humans. Science Magazine Nov 14, 2018. https://www.sciencemag.org/news/2018/11/massive-crater-under-greenland-s-ice-points-climate-altering-impact-time-humans.

Womack, W. R., & Schumm, S. A. (1977). Terraces of Douglas Creek, northwestern Colorado: An example of episodic erosion. *Geology, 5*(2), 72–76.

Worobey, M., Gemmel, M., Teuwen, D. E., Haselkorn, T., Kunstman, K., Bunce, M., Muyembe, J. J., Kabongo, J. M. M., Kalengayi, R. M., Van Marck, E., & Gilbert, M. T. P. (2008). Direct evidence of extensive diversity of HIV-1 in Kinshasa by 1960. *Nature, 455*(7213), 661.

Yeung, J. (2020). A blob of hot water in the Pacific Ocean killed a million seabirds, scientists say. CNN update January 16.

Resources

Contents

M. R. Welford, R. A. Yarbrough, *Human-Environment Interactions*,
https://doi.org/10.1007/978-3-030-56032-4_5

Learning Goals

After reading this chapter, you will be able to:

- Evaluate our technological ability to continue to extract and utilize current resources and locate new resources to maintain global consumption levels.
- Evaluate whether we will be able to avert the impending resource and pollution crises that *The Limits to Growth* first predicted in 1972? Or are we facing the "Inconvenient Truth" or a "Long Emergency"?
- Appraise arguments that blame population growth for potential resource shortages.
- Explain how a shift to a US-diet led to a more resource-intensive and higher GHG emissions scenario.

5.1 Natural Resources: Their Historic Exploitation and Exhaustion

Although still highly debated today, Easter Island's collapse (see Extinctions chapter) appears to be a combination of isolation, destruction of their natural resource base (e.g., soils, forest, and birds), introduction of an alien species—the rat—and political misadventure. The collapse of Viking Iceland appears to somewhat mirror Easter Island, in that the Vikings cut all native forest, burned grass sod for fuel, and, during unfavorable climatic episodes, ceased to trade with Europe. In both cases, their consumption of natural resources, and the subsequent failure of their associated ecosystem services which were absolutely necessary for life, continued right up to their collapse. Yet these brutal Malthusian events in Easter Island and Iceland are likely to happen again if we continue to exploit and exhaust our natural resource bases. We must alter our trajectory! But before we can respond, we must identify what natural resources are, how we utilize them, and how their utility varies across time and space.

> Resources have utility, and natural resources are those that are derived from the earth/biosphere/atmosphere and exist independently of human activity. A resource does not exist without someone to use it.

Different individuals or groups value resources differently. We describe this as "their environmental cognition" which can be defined as the mental process of making sense out of the environment that surrounds us. Several factors affect cognition, these include cultural background; view of nature; social change, gender, ethnicity, education, and income; scarcity; and technological and economic factors. For instance, the La Brea Tar pits were a nuisance to early pioneers, but today represent the second largest reservoir of oil in the lower contiguous 48 US states. We can classify natural resources into those that are perpetual such as the sun, wind, ocean waves, and tides, those that are renewable such as forest timber, and those that are nonrenewable such as oil or other minerals. In addition, there exist potential resources such as deuterium, an isotope of hydrogen that will be critical to the future of fusion nuclear technology. These are potential resources that we know have utility, but we do not have the technological ability and/or economic incentive to utilize yet. Some natural resources are difficult to classify such as groundwater and soil, and although renewable, these are renewable on time scales that are beyond human use.

Among the many factors that affect whether society exploits, conserves, or preserves natural resources is the conflict between a nature-centered view of resource use and a human-centered view of resource use. All individual organisms, species, and ecosystems have inherent value in a nature-centered view, and at its most extreme manifestation, all nonhuman entities have a right to exist and killing them is immoral. Some peoples and religions extend this to nonliving entities such as mountains and rivers; for instance, the river Ganges is sacred to all Hindus in India. In a more recent context, in a sense, designating a cultural or natural location as a UNESCO World Heritage site creates a location that has inherent value and should be preserved at all costs. This comes with a litany of problems—the creation of the Nanda Devi and Valley of Flowers National Parks as a World Heritage site led to the exclusion of all people from these locations, including native local peoples who lost their land and sources of income including herding, mountain guiding and arable croplands, access to their spiritual heartland, and their identity. In contrast, in a human- and/or business-centered view, the quality of human life is the priority and all other nonhuman entities' needs are relegated to a secondary role. In other words, we concern ourselves with the economic value of a resource or species and not its inherent value. For instance, Japanese businessman, Kiyoshi Kimura, paid 1.38 million euros for a single bluefin tuna at the Tsukiji Fish Market in 2018. And then on Jan 4th, 2019, he reset the record at 2.7 million euros for a 278 kg fish caught off the northern coast of Japan.

Today, it is broadly accepted that economic activity must account for the environmental costs of production. This is frequently identified as having a sustainability consciousness (Cutter and Renwick 1999). In the past, lumber supplies were easily identified with profits and numbers of employed peoples. In contrast, today, lumber production is discussed in terms of accelerated soil erosion, downstream water quality and dam and reservoir siltation, atmospheric pollutants released by lumber and pulp mills, and the value of intact forests to other industries, particularly recreation and tourism.

5.2 Geographic Scale and Resource Consumption

According to McMichael, four great periods characterize human-environmental relationships in the Anthropocene. Each has borne witness to the evolving nature of human-environmental resource consumption, threshold exceedance, and connectivity between fellow humans through trade and transportation that dictate such things as disease and cultural and industrial innovation transmission, trade goods flows, and migration of peoples. Human-environmental impacts have evolved from local to global effects, as technological innovations and population growth have provided humans with the ability to cultivate more land, cut more forest, extract more minerals, inhabit more areas, kill and consume more animals, and pollute more of the earth. To put our existence in context, according to Damian Carrington, 83% of all terrestrial wild animals, 80% of all marine mammals, 50% of all plants, and 15% of all fish have perished because of human activities.

The first epoch of the Anthropocene occurred 5000–11,000 years ago, depending on where and when humans first adopted sedentary living in permanent settlements, cultivating the land and domesticating plants and animals. In Eurasia, this transition from hunter-gatherer to farmer occurred, according to Lazaridis and coauthors in Asia, approximately 11,000 years ago in two distinct locations: one in the southern Levant that includes modern-day Israel and Jordan, and the other in the Zagros Mountains of western Iran. These two populations mixed with early farmers from another slightly later hearth of agriculture in Anatolia, and thereafter, farming spread westward, from Anatolia into Europe, from the Levant into East Africa, and from the Zagros Mountains into the Eurasian Steppe; later still, Iranian farmers and Eurasian steppe pastoralists spread agricultural innovations into Southeast Asia. At this point, human impacts remain local, dispersed, and minimal. Humans still lived in a large world, where coupled human networks were limited. Although humans advanced as a wave across the Eurasian continent and on into Australia, connectivity amongst humans and the sharing of cultural facets and technological innovations in hunting, agriculture, and architecture progressed in a very on-again off-again and spatially haphazard manner.

Although the advent and spread of both pastoralism and farming can be identified in the geologic record, a necessary requirement that defines the Anthropocene, it should not be forgotten that pre-agricultural humans prior to 11,000 years ago had, through *associated blitzkrieg overkill*, wiped out much of the megafauna (those animals over a 100 kg in weight) from Eurasia and Australia as humans advanced across these regions. However, the Australian megafauna extinction does seem to have been delayed until 30,000 years after the first presence of humans in Australia. Only where humans coevolved with a megafauna in Africa does a mass megafauna *overkill* appear delayed until very recently.

The second epoch of the Anthropocene occurred, according to Watts and Strogatz, between 1000 and 3000 years ago when we began to trade across continents. First Alexander the Great and then the Silk Route united east and west Eurasia. The full integration of Eurasia was complete when, in the 1330s, plague erupted in Central Asia, and by 1347, it arrived in Europe carried by Genoese and Venetian traders. These same traders transshipped silks, ivory, and gun powder and gun technology from China to Europe.

But it was the Greeks and then the Romans who first broadly impacted the earth by radically altering the landscape and hydrology of Italy, southern France, Greece, and Spain. The Romans cut millions of hectares of Mediterranean forest, replacing it with large-scale farms of citrus fruits, olives, grapes, and millions of sheep and goats. Roman forest removal reduced evapotranspiration across the northern Mediterranean, warming, drying, and desiccating the region. The Romans also engineered many of the streams and rivers for agricultural and urban use, and here we see, for the first time, significant, near industrial-scale mining of surficial and near-surface mineral deposits that triggered severe local deforestation and consequently accelerated soil erosion. Roman road construction to facilitate commerce, trade, and mineral extraction and to provide rapid reinforcement of the Roman frontier with soldiers also opened up the landscape as forests adjacent to roads were cut to reduce the likelihood of traders and soldiers being ambushed. Roman deforestation restricted initially to the Mediterranean spread northward, with the Roman conquests of northern Europe. The opening up of the European forests resulted in an irreversible openness of the European landscape. As the forest was cut by Romans, we see the extinction of the aurochs—the ancestor of the modern cow—and the isolation of wolves, beaver, bison, and bear, among many other species, to the periphery and mountainous regions of Europe. We also saw the Celts squeezed into marginal, peripheral landscapes on northwestern shores of Europe in Brittany, Normandy, Ireland, Wales, the Isle of Man, and Scotland.

The third epoch of the Anthropocene began in 1492, as humans became intercontinental travelers, with Europeans finding and colonizing the Americas, Australia, New Zealand, and other Pacific islands. The commercial exploitation of these new lands by white colonial landowners and their convicts was substantial and detrimental; for instance, 50 million indigenous Americans died due to the introduction of flu, measles, and smallpox. The subsequent American slave trade from West Africa into Brazil, the Caribbean, and North America is a very sad legacy of this holocaust. However, widespread reforestation and associated CO2 sequestering and global cooling in the form of the Little Ice Age occurred in the immediate aftermath of this holocaust. However, the expansion of the Europeans around the world heralded the beginning of the sixth mass extinction, first observed across Atlantic, Indian, and Pacific islands, and then more recently transferred to the continents (see Extinction chapter for more details; ◘ Figs. 5.1 and 5.2).

The third epoch also saw the advent of the Industrial Revolution in 1705 and the associated Agricultural Revolution that occurred between the late 1700s and early 1800s. Beginning in the West Midlands of England in 1705, and then diffusing south and eastward into Europe and reaching northern Italy by 1815, the Industrial Revolution radically altered the ability of humans to impact natural resources at local, regional, and global scales. Massive consumption of coal in Europe initiated global warming, and more recent exploitation of other coal deposits in Africa and Australia continues to contribute to GHG production and global climate change. The Industrial Revolution did not just contribute to early GHG emissions, it also launched widespread air pollution across Europe from sulfurous coal in the form of soot and acid rain. In 1705–1805, life expectancy in the UK dropped in the industrial midlands due in part to the coal-related pollution.

Today, we live in the fourth great epoch of the Anthropocene where our global transportation networks move goods, ideas, resources, technology, people, and diseases such as COVID-19 and SARS rapidly around the globe. But even within this fourth epoch, there are temporal variations in our exploitation of our environment and natural resources. Sipple and coauthors just recently showed that we can see the signature of anthropogenically induced climate change due to the emission of GHGs in everyday weather since 1992. In fact, the early 1990s were a tipping point.

5

Fig. 5.1 Congestion on the London Underground. (@ Mark Welford 2017)

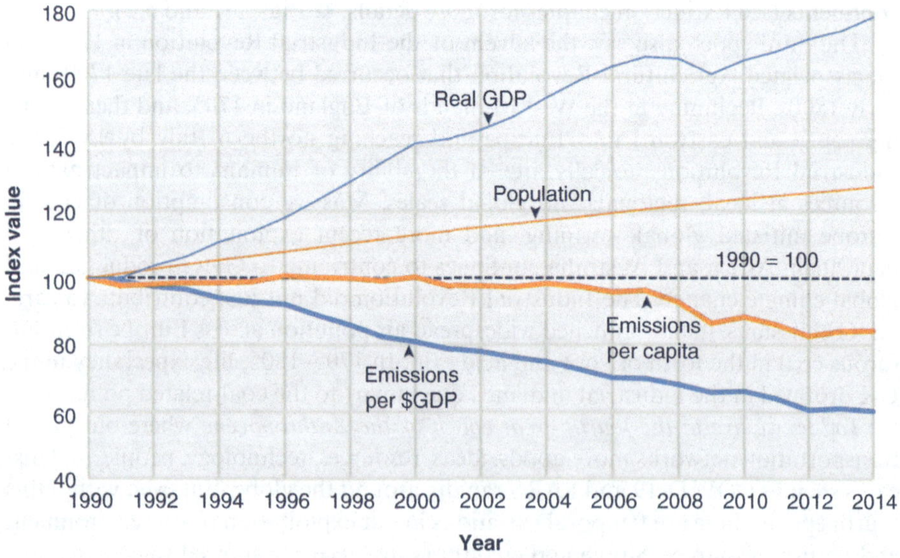

Fig. 5.2 US greenhouse gas emissions per capita and per dollar of GDP, 1990–2014. (U.S. EPA 2016)

China, following political and particularly economic turmoil initiated by Deng Xiaoping, came to embrace globalization and foreign direct financial investment as a means to feed, clothe, and educate its people. Today, China is the largest producer of GHGs and the world's largest economy. In the 1980s, China was a minor peripheral player in the world's economy. But China's growth has fueled massive natural resource exploitation beyond its borders. In order to power this economic revolution, China imports enormous quantities of coal from Australia. Australia's economic attachment to coal probably explains why several of its past and its current (2020) prime ministers have been very reluctant to publicly acknowledge anthropogenically induced climate change. As an aside, China has encouraged enormous numbers of its citizens in the last 30 years to obtain graduate education abroad while it builds the infrastructure to do this at home. In 2020, we are starting to see their success, as fewer and fewer Chinese citizens are applying for graduate education in the West.

Yet numbers are only half the story! There remains massive spatial variation in growth rates, consumption, economic development, and pollution, some or all of which have their legacy in colonialism.

The concept of overpopulation is a convenient ideology, but a fraud, as it ignores impact per capita (per person) consumption and focuses on simple numerics. In fact, the United States consumes 25% of the world's resources, yet its population of over 330 million is only 5% of the world's total population. Furthermore, the USA and Europe represent less than 15% of the world's population, but their combined consumption far outstrips most of the rest of the world. In actuality, the combined consumption of the USA, Europe, and Japan suck up 80% of the world's natural resources. What is more, France, which is five times smaller than the Democratic Republic of Congo, has about the same population. The UK is smaller in geographic size than Gabon, but has a population of 62 million compared to Gabon's population of just 1.5 million. As a whole, Africa has a population smaller than China and a total GDP which is half that of a small country such as France.

5.3 Oil and Coal and Our Carbon-Based Civilization

The reality is that our oil-based civilization will run out of its most precious resource, that is, crude oil, in the next few decades. If we do not change our oil usage, we will return to a large world, less technologically driven—one less connected as suggested by Kunstler. Intertwined with the dwindling availability of cheap energy and the exceedance in the Earth's carrying capacity is the specter of climate change. Without easy access to fossil fuels, adapting to these environmental changes will be much more difficult.

Now when we say, "run out," we are indeed referring to the eventual day when no more oil is available to be pumped out of the geological innards of our planet, but for practical purposes, we henceforth define the impending oil/energy crisis in different terms. Here, in using the phrase "run out," we refer to the day in the not-too-distant future when gaining a barrel of crude from tar sands, or shale deposits, or deepwater offshore drilling sites, or other difficult-to-access oil reserves will be

5

so cost-prohibitive that, for all intents and purposes, there will be no more oil for the masses. When oil becomes too expensive, so will all of its distillates, such as gasoline, kerosene, diesel, avgas, heating oil, and the like. People will no longer be able to afford airplane tickets because the cost input (aviation fuel) would be so expensive that only the very wealthy could afford to buy a seat. Of course, the people who work on the airlines, including the pilots, crew, baggage handlers, ticketing agents, and the like, will not be able to easily get to work anyway as their cars sit idle in the suburbs many miles away from the airport which used to be a busy hub of activity and employment. Imagine the effect that this will have on the global economy: airplanes in hangers, cargo ships moored because the cost of fuel is too expensive to be afforded, empty highways, and empty shelves as the transportation costs of worldwide-produced goods exceeds any profit from the sale of the same. Imagine the cost of food at the grocery store (as well as the availability of items). In case you had been thinking that solar power, wind power, nuclear power, or hydrogen fuel cells will solve the issue, rethink these ideas. Can you imagine airplanes running on solar energy or hydrogen fuel cells? Do you think biodiesel, heavy and noxious, will provide a solution to widespread transportation issues globally? Or perhaps that commercial ships or planes can be modified or allowed to operate somehow with onboard nuclear reactors? Even our nuclear-powered carriers have to dock occasionally to resupply and fuel-powered vehicles and vessels deliver those supplies.

Aside from the transportation disaster that awaits our globalized world in the future, there are a lot of items that humanity depends on for its daily existence that are made from crude oil. Plastics, pharmaceuticals, fertilizers, and pesticides are just a few of the ubiquitous items of daily existence that are sourced from fossil fuels such as oil. In fact, much of the world's current standard of living depends on these items for survival. The loss of fertilizer supplies and pesticides could result in severe food shortages in the future, which in turn could greatly affect a larger world population base. A lack of pharmaceuticals could lead to misery and suffering, especially among developing countries. As with oil extraction, before the last oil-based pesticides and fertilizers are produced, prices will soar inexorably as availability of the raw resource material diminishes. This will preclude access to these items by the least wealthy. Many plastics and polymers that are oil-based will also become scarcer, in turn raising the prices of the myriad low-quality imported goods produced in far-away cheap-labor markets. Therefore, the energy crisis is ultimately linked with the exacerbation of an already precarious global food supply and a breakdown in medical care.

5.4 Societal and Government Responses

Knack and Keefer suggest that societal responses regarding environmental resource base use, whether exploitative or preservative, are extremely important and are in large part dependent upon a country's environmental economic policy. Political responses to these environmental issues are extremely vital to the lives of the citizens within the state and district, and with increasing interconnections among societies over the last few centuries, it is clear that now, and maybe more than ever, economic policies of one country have an effect on other countries around the globe. These responses begin primarily at the top with district and country leaders who have been given the responsibility to make intelligent decisions for the future of our society, while keeping sound current economic growth policies and avoiding political suicide. Before analyzing specifics of societal responses, it is important to understand that not all decisions made by all have the same impact and power. Wallerstein asserts that a powerful set of core countries consisting of both former world hegemons and newly emerging societies have primary control and "exploit" periphery and semi-periphery countries politically, economically, and, most importantly within the scope of this text, environmentally. Three separate responses in terms of economic policy have been outlined within this section, two examples come from bona-fide core countries and another deals with a growing semi-periphery country looking to expand and become a global power. In cultivating an understanding of the current world-system and the relationship between countries, we can further assess future global policies and see how they might affect each country differently.

Data from the IPCC and U.S. EPA indicate that the total CO2 emissions from the USA have declined over the past ten years, but the nation is still among the top two in the world and the highest in terms of per capita production.

Much of the US and world decline in CO2 emissions can be explained by a sluggish economy since the global recession of 2008–2009. Lessening our dependence on oil will be critical to promote sustainability, conservation, and green economic development. While the USA is consistently lambasted for a lack of any green environmental policy and ranks in the low-to-no growth section of the Yale Environmental Performance Index ranking, there is a model that the USA could follow—either Germany or France. Both are rich, core countries with strong economies, solid environmental policies, and green public awareness. Germany, in particular, is renowned for its renewable energy production. In 2017, 36.1% of Germany's total electric output came from renewables, and according to Amelang, on January 1st, 2018, 100% of its electrical needs came from renewables. However, in the 1970s, environmental awareness in Germany, just like much of the rest of Europe, was poor. In the 1970s, Theil noted that Germany's Rhine river was described as a cesspool, and the West German government refused to cut sulfur dioxide emissions or cap ozone-depleting hydrochlorofluorocarbons. However, German environmental policy changed in the 1980s in response to the disasters of Chernobyl's numerous toxic spills into the Rhine, acid rain, and a growing environmental movement led in parliament by the world's first major Green Party. So, although not ranked at the top of the EPI rating (Switzerland), Germany is first among countries that is making itself green by design.

Shifting diets, more domestic animals, and more GHG emissions

Today, industrial-scale farming has evolved to provide us with a Western-style diet rich in calories, full of protein, and stuffed with animal by-products (meat, cheese, eggs, butter, milk). In fact, 96% of all animals living today are domestic livestock. As a result, we are consuming more resources and producing more GHGs per food calorie obtained than ever before. This global transition to an American-style diet is not sustainable. In addition to the resources (oil, electricity, water, land) used, land use changes necessary to sustain this transition into the future, GHGs emissions, and groundwater pollution, human health will suffer through more cancer and more heart-related and circulatory problems. Although richer, humans may end up dying younger, with more quality-of-life impediments, while suffering greater medical costs.

We need to reduce the overconsumption of calories and reduce food waste. For instance, a third of all people worldwide are overweight (with a body mass index of 25–29.9), while a third of all US citizens are obese (where a BMI is in excess of 30). As for food waste, globally 1.3 billion tons of food is wasted each year or a third of total food production. This amounts to US$680 billion in the Global North and US$310 billion in the Global South. According to Gunders, the average US household throws away $2200 of food each year or 1250 calories per person per day, and 90% of this is thrown away too soon. As a result, we waste 2.6% of all US GHG emissions, 21% of all US agricultural water usage, 19% of all US croplands, 18% of all farming fertilizer, but contribute 21% of US landfill content and waste US$218 billion (see Agriculture chapter 7).

We need to reduce the production of GHGs associated with the production and transportation of food. We need to reduce the distance food is shipped—the US food is typically shipped over a 1600 km. We need to eat more locally produced foods and support more farmer's markets. We need to eat more organic food and more hydroponic produced food. We need to reduce the overconsumption of animal-based protein foods; in particular, we need to reduce the production and consumption of beef, which is two-and-half to three times more polluting and more resource-intensive than other means of food production (Fig. 5.3).

▸ Example

In 2009, Ranganathan and coauthors found that 75% of all agricultural land was used in the production of animal-based foods and in the process contributed 66% of all agricultural GHG emissions, yet contributed only 37% of all protein consumed; much of this is due to embracing the American-style meat-product-based diet. In fact, they suggest that in producing the average American diet, land use and GHG emissions are nearly twice the global norm. Nearly half of the US land use and GHG emissions are due to the production of beef. They also document that for each new person embracing an American diet, a one-time release of 300 tons of CO_2 occurs for each hectare of land converted to a resource-intensive predominately animal protein diet. If we could shift people from an American diet to a more plant-based diet typical of India, Ranganathan and coauthors argue we would observe agricultural GHG emissions decline between 15 and 35%, convert between 90 and 640 million hectares of resource-intensive agricultural land use back to low-resource input land use, feed 10 billion in 2050, and reduce GHG emission from land-use conversion by 168 billion tons. ◂

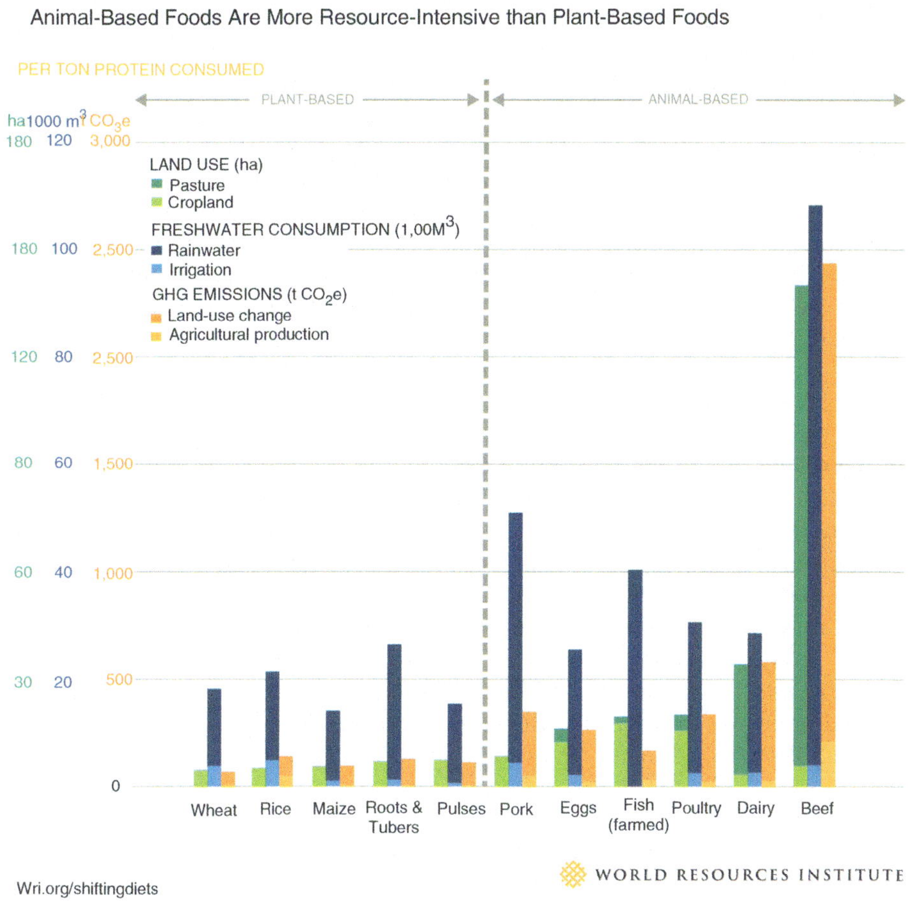

Fig. 5.3 Animal-based foods are more resource-intensive than plant-based foods (▶ wri.org/shiftingdiets). (Ranganathan et al. 2016 CC)

5.5 Economic Development and Sustainability

Recurrent crises are a fundamental phenomenon of biological existence. Whether those crises are organismally determined (as in economic or resource-based crises) or naturally determined (as in natural disasters or non-anthropogenic climate cycle shifts), living organisms have two choices: adapt and conquer, or acquiesce and die. The scales at which crises affect biological life range hierarchically from the individual-level (whether a particular organism will survive, or even thrive, in a crisis), to the cohesive (corporate) scale, to the community level, and finally to the civilization level. History has provided many examples where units at each of these scales survived or even prospered when faced with crises, while other units have failed and passed from this earth. Who survives crises, be they economic bubbles, wars, disease, or environmental catastrophes? Using these examples, we explore what they tell us today at the start of the twenty-first century about our potential to survive the coming food, oil, environmental, and climate crises. The problems we face today are not new. Beginning with the first Earth Day in 1970 and continuing with Al Gore's polarizing *An Inconvenient Truth* and Thomas Kunstler's *The Long Emergency*, the Western world is certainly aware of the energy/climate/food/environmental crisis that threatens our very existence (Gore 2006; Kunstler 2005). Getting through this crisis, however, is going to take coordinated planning, be it at the household, business, community, or civilization level. At all levels, a shared vision and a series of coordinated actions will have to occur, more than likely stemming from the thoughts and actions of either individuals or small groups of like-minded individuals.

Mapping the effects of climate change, Maplecroft suggests that poor countries in the Global South, particularly in Central and Northwestern South America, sub-Saharan Africa, and South and East Asia, are most at risk from climate change. Yet many of these countries are the least able to cope with the environmental changes that will happen. Given the already precarious state of the world food production and safety network, a drought in a prime growing region (many of which are shrinking in size) or in an area in which drinking water is sourced at the surface from lakes and streams could result in serious food and safe water shortages. The area in which vectors of disease (insects, rodents, etc.) currently live may expand or change in size as the climate changes, resulting in disease outbreaks in formerly disease-free areas.

Water Resources

Typically forests and offshore mineral rights are owned by governments, but air and water are considered a *common property*; as a result, these are exploited by private individuals and corporations. Let us not forget that while the costs of exploitation are shared among all owners of the resources, the benefits accrue to the individual user, and as a result, it is in the individual's best interest to increase exploitation or pollution of a *common property*. In the USA, several classic examples of common property issues are functioning today. The draw-down of the Ogallala Aquifer in the Great Plains of the USA is a prime example: multiple private (i.e., rural homeowners with wells, farmers, oil companies) and public entities (towns and cities providing drinking water) tap this renewable resource, but at rates that are not sustainable. The use of the Colorado River flow by states, municipalities, and famers means that a dribble of water flows across the US-Mexican border into the Gulf of California; as a result, a once rich biodiverse delta is dead!

Common property issues are especially common within seas and oceans. There are three fundamental issues with seas and oceans: (1) living and nonliving resources (within seas/oceans) are unseen, unmeasurable, and unaccountable; (2) oceans act as the ultimate diffuser and sink for nature products of erosion and human-made pollutants; and (3) ownership and control of these water bodies are very contentious. The EU had to step in to save the Mediterranean Sea and issue rules, penalties, and offer funds to clean up sewage disposal among coastal communities of France, Spain, Italy, Greece, and several Balkan countries. Today, the northern Mediterranean is quite clean, yet sadly the southern Mediterranean is still quite polluted as most of these adjacent countries lack the funds to modernize sewage disposal. In 1970, 9% of the world's protein was obtained from oceans; today, approximately 16% of protein is obtained from oceans, although Global South nations obtain greater than 20% of their protein from marine resources. Today, the leading fishery nations are China, Peru, Japan, Chile, and the USA, with Japan, China, and Russia obtaining most of their fish from outside their territorial waters.

The 1975 Cod Wars between the UK and Iceland illustrate the degree countries will go to protect "their water resources" (see Chap. 1). From an Icelandic point of view, this is completely understandable given that foreign fisheries destroyed cod, haddock, and flounder fishing off the east coast of Canada in the 1990s, with the loss of 40,000 jobs. Both mackerel and herring fishing survive only under strict fishing limits in the North Sea. While the anchovy, swordfish, and marlin fishery off the west coast of Peru and Chile collapsed in the 1970s due overfishing and the impact of warm-water El Niños. Rather surprisingly, according to the UN, many northwestern European countries, in addition to many North African and Central Asia countries, are even now suffering water stress and are vulnerable to many further climate change that negatively affects their access to renewable water resources.

Case Study—Kissimmee Lake Chain

Multiuse management of water resources is typically very complicated and difficult to arrange. For instance, Central Florida has some of the world's most famous lakes for sportfishing. In the 1960s, the state fish and wildlife agency (now the Florida Fish and Wildlife Conservation Commission) opened an office in the town of Kissimmee to manage this world-class fishery in the Upper Kissimmee River Chain of Lakes—notably Lake Tohopekaliga (Toho) and Lake Kissimmee further downstream. Office staff took a holistic approach to management, placing high value on maintaining quality habitat for largemouth bass, other sportfish, and the communities sportfish depend upon. By the mid-1960s, the US Army Corps of Engineers had built water control structures on these and other chain lakes for flood control and land reclamation purposes. This altered the historical filling and emptying of these natural-basin lakes over time, which varied up to 3.2 m; since 1964, water levels have been maintained within a narrow 0.9 m range. This alteration in water levels eliminated the natural control of aquatic vegetation communities, fostering the prolific growth of species that soon reached noxious levels. Concurrently, accelerating human population growth in the watershed, which included the southern portion of the Orlando metropolitan area, resulted in the ever-increasing discharge of wastewater treatment plant (WWTP) effluent into Lake Toho. This huge influx of nutrient levels exacerbated the effects of limited water-level fluctuations, further fueling exuberant vegetation growth and compromising littoral-zone habitat quality where sport-fish spawn. Fisheries staff began implementing periodic drawdowns on Lake Toho beginning in 1971 (coincidentally, the year Disney World opened, itself partially in the upper Kissimmee watershed). During drawdowns, managers partially drain lakes for the purpose of controlling the growth of undesirable concentrations of near-shore vegetation and aiding in consolidating shallow-water habitats. But it soon became apparent that drawdowns were not enough to conserve sportfish spawning habitat; something had to be done to control massive nutrient loadings. The littoral zone of Lake Toho had become so overgrown with living and decaying plant muck that broad areas of former sportfish spawning habitat were no longer available. The situation prompted biologists to couple a muck removal project with their 1987 periodic drawdown; 165,000 m^3 of organic material was mechanically removed, restoring 136 ha of spawning habitat. By 1986, four WWTPs discharged 110,000 m^3 daily into the upper chain. Working diligently with local, state, and federal partners and a conservation-concerned public, fisheries staff was able to implement no-discharge rulings by state and federal agencies for these WWTPs (the effluents were primarily diverted for golf course irrigation). By 1988, nutrient concentrations in the lake were reduced by 75 to 92% from 1980 levels. Conservation of prime habitat in the Kissimmee Chain of Lakes was accomplished with the farsighted vision of fishery biologists, who helped support a billion-dollar state-wide bass fishing industry [in ~1986 dollars]. What this case study illustrates is that state government and local fishermen can take on federal agencies and multinationals and win big and save critical habitat.

It would be remiss if we did not discuss the loss of water resources such as Owen and Mono Lakes in the USA and the Aral Sea, and the negative downstream impacts of dams. Owen and Mono Lakes are casualties of California's insatiable need for water, while the Colorado River was dammed for hydroelectric power and agricultural irrigation. All three have ended up nearly dry. Today, Owen and Mono Lakes and the decreasing Salton Sea are major contributors of highly toxic salt dust to the states of California, Arizona, and New Mexico. This is true to the east and downwind of the Aral Sea as well. In all three cases, the dust is contaminated with pesticides and fertilizer, and additionally heavy metals downwind of the Aral Sea. To reduce Owen Lake's dust, the dried-up lake bed is now irrigated with a sprinkler system that consumes more precious water but has seemed to solve its dust problem. According to Mansur Mirovalev, some of the salt dust from the Aral Sea can be traced to Greenland and Japan.

Dams necessary for sustainable hydroelectric power (HEP) production are themselves quite destructive to the environment. One of the most researched and contentious examples is the Glen Canyon Dam on the Colorado River at the head of the Grand Canyon. Although the damming of Hetchy Valley just to the north of Yosemite National Park initiated the modern environmental movement in the USA (see later in the chapter), the Glen Canyon Dam completely changed the downstream hydrology of the Grand Canyon. The Colorado River water is now colder (water is released from the cold, bottom layers of Lake Powell), and this prevents humpback chub from spawning below the dam. The reduced floods lead to significant erosion of flood-bars, once campsites for rafting tour companies, and has meant even human waste has to be removed from the canyon where seasonal flooding once did that job! Although periodic releases of high water from the dam now occur, the natural ecology of the canyon has yet to fully return.

▸ **Example**

In 1960, the Aral Sea covered 67,000 km^2 and was the fourth largest inland body of water. It supported a significant fishery, but in 1965, the two rivers that maintained its volume were increasingly diverted to irrigate 5–7.5 million hectares of cotton fields which were part of the Soviet Union. According to Micklin, the 1960s annual water deficits into the sea were ~12 km^3 per year; however, this increased during the 1970s and 1980s to ~30 km^3 per year. As a result, the sea rapidly decreased in volume, increased in salinity from 10 g/l to over 100 g/l of salt, became too toxic for fish resulting in the collapse of the fishery, split into two deltaic reedbeds (homes to hundreds of species of plant, birds, insects, and mammals), declined from >500,000 to just 12,000 hectares, and initiated deflation or wind erosion of the salt dust from the dried seabed. The salt dust laced with pesticides, fertilizer residue, and heavy metals contributes to infant mortalities downwind of the sea in excess of 100 deaths per 1000 births, some 7–10 times the US levels. Recent efforts to save the Small Aral Sea are underway, and a small fishery catches in excess of 1000 tons of fish per year, although the toxicity of the fish is quite high. ◄

5.6 Nonrenewable Versus Renewable Energy and Climate Change

Coal

In 2015, coal still provided 24% of the EU's electricity from some 280 coal-fired power stations, while contributing 18% of the EU's GHGs. Air pollution in the form of fine particulate matter (or $PM_{2.5}$) or soot from these 280 coal-fired power stations contributed to 22,900 premature deaths, 11,800 cases of chronic bronchitis, and over 21,000 hospital admissions in 2013. Soot is a product of sulfur dioxide and nitrogen dioxide coal-fired plant emissions. This soot causes in excess of 60 billion euros per year in health costs in the EU. This does not include health costs associated with the release of coal slag or CO_2. These health costs are also not covered by the coal industry; rather, they are paid by individuals and by governments. These costs do not include lost or reduced productivity due to poorer health. Sulfur dioxide also causes acid rain that negatively affects agricultural productivity and, through acidification of lakes, reduces fish stocks. SO2's most immediate effect, however, is to cause widespread asthma and other chronic respiratory illnesses. Nitrogen Dioxide causes eutrophication of water bodies and seeds ozone and smog, causing respiratory irritation, chronic obstructive pulmonary disease (COPD), and cardiopulmonary disease. Mercury is also a by-product of coal burning and can trigger heart disease and impair learning among children. Approximately 1.8 million EU children exhibit mercury levels above WHO safety limits. The majority of this pollution comes from 30 toxic power stations, most of which are located in Germany (8 plants), Poland (6 plants), and the UK (5 plants). And much of this pollution moves eastward on westerly winds, contaminating Romania, Bulgaria, Greece, and Turkey. These 30 toxic power stations also produce nearly 50% of all GHGs emitted from European coal-fired power stations. Coal-fired stations across the EU must reduce emissions by 8% per year if global warming is to be limited to just 2 °C by 2040; however, coal-generated GHG emissions across Europe are declining at 2.3% per year (see Jones et al. 2016 for more details).

There is some more good news: since 2007, the USA and Europe have halved coal-powered electrical generation. According to Ambrose and Goodly, in 2019, Europe decreased coal-powered electrical generation by 25%, and the USA by 16%. Across the globe, power generated from coal plants dropped by 3% in 2019, and GHG emissions by 2%, even though China continued to expand coal-generated electricity. Most of this drop is attributed to the availability of cheaper and cleaner natural gas.

Fission Versus Fusion Nuclear Energy

Although both Chernobyl and Fukushima offer terrible lessons in pollution and loss of life when fail-safe systems indeed fail, for the most part fission-based nuclear power stations are very safe and offer significant savings in GHG emissions versus other carbon-based power stations. For instance, Sizewell B nuclear power station has produced only 6.04 g CO2e/kWh over its lifetime including its construction. Typically, according to Sovacool, across the world, from mining uranium to enrichment to waste storage, some 66 g CO2e/kWh is emitted. Sizewell B supplies up to 3% of all UK homes with low GHG emission electricity; in fact, it saves 3 million tons of GHG emissions each year versus a coal-powered electric-generating power station. Over 40 years of operation, that will amount to a savings of 120 million tons of GHGs.

Fusion energy, on the other hand, has continued to prove difficult to implement. Numerous sources suggest we will have fusion-generated electrical energy input into our electrical grids sometime in the 2030s. At this point, humanity and our power-hungry civilization might be saved! In the meantime, we need to maintain what fission nuclear power we have, reduce our use of coal- or oil-generated electricity, and invest in solar, wave, HEP, biomass, and wind-generated electricity.

Renewable Energy

Renewable sources of energy are those sources that are not finite or exhaustible, but some are flow-limited. These include solar, wind, wave, geothermal, biomass, hydrogen, and hydro-electric power (HEP). To illustrate with solar energy, flow-limited simply means that the total amount of sunlight hitting solar panels is limited by the maximum amount of sunlight incident at the top of the atmosphere and that amount of sunlight intercepted as the rays pass through the atmosphere. Dust particles, water vapor, clouds, ozone, and smog all reduce the amount of sunlight hitting the earth's surface. Just as in all forms of energy production, energy-production efficiency also affects the total production of energy. Solar energy efficiency is typically 23% (today, solar panels only convert ~23% of all solar radiation incident on the panels into electricity); wind turbines are 50% efficient; wave power is thought to be able to yield 40% efficiencies; geothermal yields 12–25%; biomass is 75–80% efficient in direct heating, but only 20–25% efficient when generating electricity; hydrogen fuel cells driving electric motors exhibit efficiencies on the order of 40–60%, which is twice internal combustion engine efficiencies; and finally HEP exhibits ~90% efficiencies.

Hydrogen power is compromised by the high input cost to create hydrogen in the first place. HEP power is limited to countries with abundant water resources (either rain or river flow in the case of Egypt) but is also limited by the very high initial costs of dam construction. However, the International Hydropower Association suggests that in 2018, China with 8.54 GW, Brazil with 3.87 GW, Pakistan with 2.49 GW, Turkey with 1.09 GW, Angola with 0.67 GW, Tajikistan with 0.61 GW, Ecuador with 0.56 GW, and India with 0.54 GW are the fastest growing countries for new installed HEP capacity. Indeed, Ecuador (with loans from China) generated 58% of its electrical needs through HEP in 2017. The utility of wave power is limited because only 2% of the world's coastlines exceed a power density of 30 kilowatts/meter necessary to generate electricity from waves. The USA, Canada, the UK, Ireland, Norway, France, Australia, New Zealand, South Africa, Chile, and Argentina have wave powers that typically exceed 40–60 kilowatts/meter! This clearly provides advantages to Global North countries and relative disadvantages to tropical and Global South countries. Wind power is limited, with again mid-latitude areas benefiting from their location coincident with the maximum transfer of energy from tropical regions to polar areas. In other words, mid-latitude areas are the windiest places on the planet and hence have the highest potential for wind-generated electricity. But even in these locations, some places are windier than others: Iowa is the windiest state in the USA and has a high density of windmills, while many coastal areas are also quite windy.

In contrast, solar electrical generation does not appear biased against Global South countries; quite the opposite, the Sahara, the Namib, the Gobi all have few cloudy days. Even the UK, with its high latitude and cloudiness, is the third largest producer of solar power in Europe, yet this is still a paltry 6% of total UK electrical production. However, combined with wind (20%), biomass (12%), and HEP and wave, renewables provided nearly 40% of UK electrical needs in late 2019. That year, for the first time since the Industrial Revolution, renewables exceeded nonrenewable energy production in the UK. Elsewhere in Europe, according to Sophie Vorrath, Portugal produced more than 100% of its electrical needs for March 2018 from renewables, while, on average, it produces 62% of its electric demands from a combination of HEP, solar, and wind.

Because defraying initial investment costs are important to many countries as are the need for portability, high-power efficiencies, and the need to have suitable energy conditions, wind and biomass, coupled with solar electrical production, offer the greatest bang-for-the-buck for the Global South. The cheapest and easiest to develop are wind and biomass. Converting to renewable electrical energy production should also be coupled with a move to electric vehicles, and here Norway leads the way. Today, Norway has the largest number of electric vehicles per capita—this was achieved through tax incentives.

5.7 Conservation of Natural Landscapes and Ecosystems

Saving Ecosystems

Attempts to save ecosystems and hence stop multiple extinctions are political minefields prone to political whim and fancy, and costlier and more difficult, as saving an ecosystem is harder to sell than a cute endangered animal. Saving broad, spatially contiguous areas that harbor rich or unique ecosystems is especially difficult, given competing designs for the land and the fact that all ecosystems are three-dimensional in nature, including the land surface—its flora and fauna, the subsurface, and the atmosphere above.

Amongst the two most critical landscapes we need to conserve are the tropical lowland rainforests and the boreal forests of North America and Russia. The boreal forests are being cut for commercial timber, whereas the tropical rainforests are subject to various pressures including commercial timber production, mineral prospecting, conversion to low-intensity small-scale farming adjacent to roads, and conversion to intensive plantation agriculture including organic soybeans, maize for ethanol, sun coffee, heart of palm, palm oil, and beef and milk production, among other agricultural products (see Agriculture chapter). Yet, the Amazon, Congo, and SE Asian rainforests (that are scattered across the islands of SE Asia) are the most important terrestrial sites for carbon sequestration across earth. Many ideas have been promoted to protect these tropical forests: these include land reform, provision for pharmaceutical prospecting, creation of parks, encouraging industrialization, and expanding ecotourism.

Land Reform

In much of Colombia, the best agricultural lands are located in valley bottoms and are mostly employed for raising cattle by rich landowners. As a result, very few people live and work on the land. In contrast, large numbers of poor, disenfranchised people live on adjacent slopes and struggle to grow sufficient food for themselves and families on the poorer quality, more erodible valley-side soils. A much more sustainable approach would be to restrict cattle ranching to the valley slopes, through land reform, which is much more the case across Europe. The good soils in the valley-bottoms could support larger populations and produce significant food surpluses. The problem facing this type of land reform is that the few rich landowners control much of the local wealth and can influence local, regional, and national legislators.

Conserved tropical forests can provision the world's pharmaceutical industry

Tropical forests are vast repositories of undiscovered medicines and crop plants. For example, the anti-parasitic veterinary drug Ivermectin was harvested in 1975 by Merck from a soil microorganism located in Japan. According to Crump and Omura, Ivermectin has sustained $1 billion in annual sales for the last 20 years. With this in mind, Merck & Co. Inc. instituted a $1 million partnership with Costa Rica's National Biodiversity Institute (INBio) in the late 1990s to locate additional drugs. Nevertheless, most drugs are synthetic and little forest is needed for serious prospecting.

Parks

Across tropical countries (i.e., Brazil, Malaysia, and Indonesia), governments are increasingly reluctant to commit land to protection when they know the land might be needed by hungry citizens and rich cattle ranchers in the near future. When pressured to explain why tropical rainforests are being cut, politicians such as Brazil's President Bolsonaro respond that parks benefit foreigners, but the needs of our people are greater. Furthermore, tropical parks tend to be in remote frontier regions that justify limited investment to protect the land. Their remoteness also ensures ecotourism is minimal. As a result of the lack of government investment and limited private ecotourism investment, most tropical parks are inadequately staffed.

▶ Example

For instance, Brazil has 30 nature preserves in the Amazon that, according to Rauch, employ just 23 guards. In contrast, the US National Park Service employs over 4000. Rauch also notes that these guards are unarmed and all 30 Brazil's Amazonia parks have poaching problems, 23 have been invaded by squatters, logging activity affects 18, land titles are disputed in 18, eight harbor indigenous populations, and only 12 parks lie completely within public domain. ◀

All of these parks offer ready access to encroachers. Ecuador also struggles with many of these issues—the eastern portion of Podocarpus Park is deemed off-limits to ecotourists because of the presence of gold miners in the park, while the private Bosque Protector Cerro Blanco near Guayaquil has struggled with land squatter incursions (supported by local politicians) on its northern border in 1997, 2000, and 2010. Yet, Cerro Blanco is an important conservation area for the great green macaw.

According to Cutter and Renwick, the typical scenario that tropical parks face is one of progressive degradation. This progression is as follows: populations build around boundaries, roads are built, prime timber resources are quickly exhausted around the edge of park boundaries, and yet the park remains a magnet for loggers. If mineral deposits are discovered, then miners move in (e.g., Podocarpus NP). When perceived illegal activities can be carried out with impunity, then the full-scale assault begins. Finally, once hundreds or thousands of peasant families have helped themselves to land and cleared forest, it is nearly impossible to evict the land squatters.

Industrialization

In 1950–1990, forest cover in Puerto Rico (PR) increased from 9% to 37%. During this period and since, more land has reverted to forest in PR than in any other country. The Tandayapa and Mindo Valleys west of Quito, Ecuador, have also seen substantial reforestation due to ecotourism investment (see later case study). Most forest in PR reverted to forest in humid, upland, sugar- and coffee-growing regions characterized by unrestricted out-migration to the USA and populations of small landholders earning more money through off-farm sources. PR is unusual, as it is part of the USA, and so there are no restrictions to migration to the mainland USA. And secondly, investment in commercial activities in PR by mainland US investors and through remittances is not restricted. Direct US investment and remittances have created more attractive social, economic, and educational opportunities in PR cities, and this has compounded problems farmers face in obtaining farm labor and as a result caused more land abandonment. In PR, planters and farmers have also had to compete with producers from other countries who had access to more capital and better land. Thus, many PR farmers have simply not been able to compete, and so many have allowed their lands to revert to forest.

The case *against* Ecotourism as a significant form of conservation

As a means to protect forest, wildlife, and other landscapes, ecotourism is not a panacea, in part because business volume tends to be low and ecotourists tend to be very fickle. In other words, the volume from year-to-year of ecotourists is quite variable because rapid changes in public image of place (within mainstream media and social media) are possible due to perceived changes in crime, in political conditions, and in disease, among many other things. And a place might become too mainstream and many ecotourists will look elsewhere for "authentic" experiences. Ecotourism is also a double-edged sword—it tends to be lucrative for governments and foreigner-born residents who can invest sufficient capital into a venture that meets the needs of rich, elite ecotourists (who can travel abroad), but ecotourism businesses are difficult to initially capitalize among locals and indigenous peoples. Ecotourism is also limited because it needs an outstanding attraction to draw tourists, yet most national parks cannot offer this. Tropical lowland rainforests that are critical sites of biodiversity and carbon sequestration often awe first-time visitors, but these same visitors are frequently repelled by cool, deep shade, high humidity, and the myriad of dangers! Furthermore, while visitors also want to see animals, most rainforest critters are notoriously hard to see high in the canopy or only venture at night. Specifically, approximately 50–90% of life in rainforests live in the canopy and access is difficult and expensive. Canopy towers are expensive to build, tend to have a short life expectancy due to tree-fall and corrosion, and, as a result, few ecotourist facilities can afford their cost. The Canopy Tower in Panama is a little different: it is a refurbished radar station constructed by the US military in 1965, then abandoned in the 1990s, and repurposed for bird ecotourism. On the forest floor, trails tend to be narrow; guides might see birds or animals high in the canopy, while the next in line may also see it, but the fourth or fifth in line tends to miss the animal or bird.

For example, Papua New Guinea (PNG) is stunning: it is the world's largest tropical island and retains much of its tropical forest; it is home to a unique flora and fauna (birds-of-paradise (BOP), cassowaries, tree-climbing wallabies, tropical fish) and ~800 language groups that maintain many of their traditional cultural elements. Beginning with Wallace's zoological explorations in the 1800s, PNG has become the holy grail to birders across the world. Most, if not all, birders aspire to visiting PNG. Yet few birders visit this birding nirvana! Even David Attenborough's "Life on Earth" video of him being outdone by a lesser bird-of-paradise failed to help. In fact, less than 50,000 tourists visit PNG annually compared to greater than 600,000 that visit Fiji and its principal attraction—its beaches!

PNG does sustain a few very expensive, remote ecotourism facilities, but its great attractions and close rich neighbors (i.e., Australia, New Zealand, Japan, and China) fail to match Fiji. Foale and Macintyre argue that two quite separate forms of ecotourism exist in PNG and elsewhere across the Southwest Pacific: low-budget, low-impact, small-profit facilities for backpackers run by locals; and foreign-owned tourism businesses that offer comfort yet are expensive, high-profit, high environmental impact, and cater to diving, birding, and beach tourism. These stem from efforts in the 1990s to generate alternate forms of income other than extractive industries such as mining, logging, and blue-water fishing. Today, high-end, high-overhead, expensive ecotourism lodges dominate the market, while small family-run ecotourism businesses have been run into the ground as their profit margins cannot compete with compensation offered by competing extractive industries for forest resources and land. According to Sakata and Prideaux, community-based ecotourism facilities also struggle to achieve profitability in PNG because of several significant geographic factors: a poorly developed road network, a limited and expensive telecommunications network, a low rural gross domestic product (GDP) that limits local investment capital, and PNG's peripheral position to a source of rich, mobile tourists such as Americans and Europeans. Sakata and Prideaux also argue that the principle ecotourism attractions—remote hill tribes and difficult to locate birds-of-paradise—create a "periphery-of-the-periphery" situation that drives up costs. So, although cultural tourism and ecotourism offer PNG an additional source of foreign revenue, structural barriers inhibit PNG from maximizing its potential "green" income.

Nevertheless, some individuals are diversifying their income streams and developing small-scale ecotourism facilities that offer outstanding opportunities to observe the unique flora and fauna of PNG. Keki Lodge and owner and guide Moyang Okira offer, just west of Madang on the north coast, the opportunity in the remote Adelbert Mountains to see the extremely rare, endemic, and range-restricted fire-maned bowerbird, the Pesquet's or vulturine parrot, and, from hides, magnificent bird-of-paradise. Their tourist income supplements monies made from coffee and betel nuts. On Western Papua, Indonesia, the community of Malagufuk in the Klasow Valley to the east of Sorong, with help from local guide Charles Roring, offers the opportunity to see northern cassowary, twelve-wired bird-of-paradise, and king bird-of-paradise from hides.

▪▪ Ecotourism and Disturbance

Regardless of its motives, ecotourism necessarily involves disturbance to animals and damage to local human economies. Disturbance of birds by humans includes mortality changes, behavioral habituation, and related, but nevertheless disturbing, phenomena of increased soil erosion around sites of bird habitation. For instance, Mullner and coauthors found that the survival rates of hoatzin chicks are significantly lower in tourist-exposed nests. According to Walker and coauthors, Magellanic penguin chicks raised in unvisited areas fledge earlier than chicks raised in visited areas. Additionally, Forbes and coauthors found that disturbance by humans produces patches of partially or totally denuded ground and even relatively low-intensity, small-scale disturbances have immediate and persistent effects on Arctic vegetation and soils.

Sekercioglu argues that damage to local economies can be extensive and long-lasting. Isaacs, Vivano, and Ziffer found that ecotourism profits are typically transferred out of communities and from locals to foreigners when locals cannot afford to invest in luxurious accommodation demanded by tourists or control of ecotourist businesses and resources (i.e., land) is lost to outsiders. In addition, locals excluded from lands they traditionally had access to either in the form of out-right exclusion from National Parks (e.g., in India, Bosak 2004; in Tanzania, Kamuaro, 1996; in Sri Lanka, Tisdell 2003) or, as Pratt found, that where multi-use common lands that are "purchased" and set aside as private protected areas (PPAs) has created considerable local resentment. The creation of PPAs and associated ecotourist facilities can, according to Isaacs, create the illusion that sufficient natural areas are being conserved and thus encourage entrepreneurs and policymakers to destroy other natural areas. For instance, in Costa Rica, Honey found that PPAs are increasingly stripping funds from National Parks as more and more NGOs, scientists, and tourists work with or visit PPAs.

Tourism, and especially bird-related ecotourism, often involves Westerners traveling to more peripheral, isolated, and hard-to reach communities, resulting in considerable cultural and environmental degradation according to Honey, McLaren, and Tisdell. Enloe suggests tourism is politically inspired in more peripheral nations to internationalize peoples in remote communities and allow governments to more effectively control these communities. McLaren suggests degradation typically follows a pattern—forest is cleared and land converted into lodging for ecotourists, inadequate sewage disposal systems are overwhelmed, and garbage (particularly plastics) accumulates.

Case Study: Ecuador—the good, the bad, the ugly, and ecotourism

Yet in reality, conserving or preserving large-scale, spatially contiguous ecosystems is something that rich countries, that have gone through their exploitative stage in wealth generation, have the means to accomplish. This is not saying newly emerging countries do not have the will to conserve or preserve. For instance, although the Ecuadorian government has more comprehensive environmental legislation than many far richer countries, it still lacks the means to achieve many of its goals. Sadly, at times, the Ecuadorian government has to make critical choices between social welfare and environmental welfare. In Ecuador, four cases stand out: the ongoing deforestation of the western slope of Ecuador, the construction of the 2001 Oleducto Crude Petroleum (OCP) oil pipeline through the Mindo-Nambillo protected forest, the migration of thousands of Ecuadorians to the Galapagos Islands over the last two decades to cash-in on its tourism, and the 2016 opening of oil exploration of Yasuni National Park.

On August 3, 1993, Ecuadorian ecologist Eduardo Aspiazu, botanist Alwyn Gentry, ornithologist Ted Parker, and their pilot died in a plane crash while conducting an aerial survey of a remote patch of Ecuadorian cloud forest near Guayaquil as part of a Rapid Assessment Program inventory. They were there because the western slope of the Ecuadorian Andes is one of the most deforested environments on earth—only 4% of the original forest remains, yet according to Dobson and Gentry, greater than 20% of lowland western Ecuador's plants are endemic. Of these endemic plants, 4% are critically endangered, but federally protected areas are inadequate to stave off extinction: it appears that private-protected areas (PPAs) are the only hope. Much of the deforestation occurred in 1960–1980 as Ecuadorian laws encouraged the conversion of forest to agricultural activities, such as palm-oil production, to reduce rural-urban migration. Yet, sustained deforestation of western Ecuador began in the 1940s when primary lowland forest was first cut for banana production for the US market. Today, fragments of lowland, foothill, and montane primary rainforest remain, for example, 589 ha owned by Mindo Cloudforest Foundation spread out over 4 reserves; the Jocotoco Fundacion in partnership with the World Land Trust protects 1200 ha at the Yanacocha Reserve, 2000 ha at the Río Canandé Reserve, 2259 ha at the Buenaventura Reserve, and 1438 ha at the Jorupe Reserve. In addition, the Bellavista Cloud Forest Reserve protects 809 ha; the Cerro Blanco Protected Forest protects 6078 ha near Guayaquil; the 6000 ha Maquipucuna Reserve owned by the University of Georgia; and the 19,200 ha Mindo Nambillo Ecological Reserve, the largest private protected reserve in Ecuador. There are several federally protected parks in western Ecuador, but most are poorly staffed. These include the 128,000 ha Machalilla National Park; the 3400 ha Pululahua Reserve; 119,172 ha Mache Chindul Reserva Ecologica; the 149,900 ha Los Illinizas Reserva Ecologica; the 752,235 ha Cotacachi-Cayapas Reserva Ecologica; the 49,350 ha Manglares-Cayapas-Mataje Reserva Ecologica; and the 86,589 ha Manglares-

Churute Reserva Ecologica. All of the federal reserves are remote and support limited ecotourism and suffer from illegal logging, hunting, gold-prospecting, and colonization because most of these activities are carried out with impunity. Sadly, Cerro Blanco also suffers from these illegal activities along its eastern boundary as the owner, the Cemento Nacional, turns a blind eye to these activities. As a result, these reserves are experiencing progressive degradation, while local, regional, and national politicians argue that these parks benefit foreigners, not the greater needs of their voters.

In 2000, a new Trans-Andean oil pipeline was proposed to increase oil transportation capacity across Ecuador, from its oil-rich production fields in the Amazon to its exporting and refining port of Esmeraldas on the Pacific coast. The goal was to increase oil revenues and government taxable income at a time when Ecuador was struggling financially following a period of high inflation and a decade of ten presidents. Funding for the $900 million pipe was provided by a US/Canadian/German consortium, and by 2001, the pipeline had been built through several ecologically sensitive and unique habitats. However, we (the authors) found that four loosely coordinated groups of environmental stakeholders from locals to global NGOs formed a resistance to the proposed route. Although they failed to stop pipeline construction, together they helped the Tandayapa Valley, Mindo, Milpe, and Los Bancos emerge as one of Ecuador's primary ecotourism destinations, as their aggressive manipulation of local, regional, national, and international news media spread the word far-and-wide of the unique landscape and flora and fauna of northwest Ecuador. Among their successes was the establishment of Ecoruta Paseo del Quinde. Although this is the maintenance road for the oil pipeline and new electricity pylons through the Tandayapa Valley, the Ecoruta allows access to critical habitat for birders and ecotourism operators. Purchase of 950 hectares of cloud forest in 2001 by the Fundacion de Conservacion Jocotoco to create the Yanacocha Reserve to save the forest surrounding the Inca (Yanacocha) ditch was followed in 2010 by another 250 ha addition. This reserve serves to protect a strong-hold of the critically endangered black-breasted puffleg (*Eriocnemis nigrivestis*), endemic to Pichincha Volcano. One NGO, the Mindo Cloudforest Foundation, wrote two editions of Ecuador's official National Avitourism Strategy. As one local environmentalist stated, "A bad thing became a good thing… Mindo became a model for Ecuadorians to conserve nature, a model about birding, a model about how local people can work on these issues---you can ask everybody in town and everybody can tell you something about conservation." Tourism has begun to replace unsustainable agricultural activities such as cattle ranching and logging. Today, in Mindo, two butterfly farms, a chocolate farm/factory, a beehive tour company, several zip-lining and forest canopy tours, and several tubing outfitters using the Mindo River support more than 20 restaurants and 62 hotels and provide hundreds of jobs.

The imagery of pristine beaches, the blue-footed boobies, and tortoises that was used to sell the Galapagos Islands in the 1970s, 1980s, and 1990s is sadly under threat. Today, 25,000 people live on the four largest islands of Isabela, Santa Cruz, San Cristobal, and Floreana, and, according to Benitez-Capistros and coauthors, catering to the over 200,000 tourists who bring in excess of 85 million USD/year. Both their direct and indirect impacts are substantial. Once wild land is now being turned into farms and roads and being covered by buildings. Locally obtained rock, sand, and timber is being consumed as building materials, while the growing residential populations and tourists demand more water, more electricity, better sewage disposal, and more consumer goods. More tourists and the importation of more consumer goods place the islands at a greater risk for the introduction of exotic, invasive species, in addition to humans. More residents and more tourists have meant more fishing for locally consumed fish, but the increase in residents has also fueled an increase in illegal fishing of shark fins, sea cucumbers, and lobsters. Although access to the other 15 islands is limited to 74 sites, the sheer numbers of tourists pose many problems, and in a sense, the Galapagos Islands are being loved to death. Not least is the fact that much of the revenue generated by tourism is not invested in conservation efforts. Most of the tourism revenues leave the islands as most tourism operators are either mainlanders or non-Ecuadorians.

The 9823 km^2 Yasuni National Park is a paper park. It was established in 1979 to protect critical Amazonian habitats and several uncontacted Huaorani groups, and became a UNESCO Biosphere Reserve in 1989. However, poaching of monkeys by non-Huaorani groups is common, while petroleum exploration was initially banned, but that was recently rescinded. In 2007, then-President Correa, backed by the UN and conservation groups worldwide, offered not to explore for petroleum in the park if Ecuador was paid not to! Although many governments and NGOs pledged millions, only $200 million was forthcoming, and rather reluctantly, the Ecuadorian government opened up the park to petroleum exploration. However, the Ecuadorian government maintains that oil production, now in excess of 23,000 barrels/day, is minimally invasive!

5.8 US National Park Conservation and Preservation

As a by-product of the creation of Yellowstone, the fight over the use of the Hetch-Hetchy Canyon in California led by John Muir, and the efforts by Gifford Pinchot, the first Chief of the Forest Service, the USA now has five Federal Bureaus charged with administering and managing the nation's forests. These are the Soil Conservation Service, the Tennessee Valley Authority, the U.S. Fish and Wildlife Service, the U.S. Forest Service, and the Bureau of Land Management (BLM). The BLM alone is responsible for 253 million surface acres and 700 million subsurface mineral rights. Today, these Bureau's enforce the Multiple Use – Sustained Yield Act of 1960, conserving land resources following Pinchot ideals, and requiring that forest resources must be made available to the greatest number of Americans. On the other hand, following Muir's philosophies, the Wilderness Act of 1964 sought to preserve primitive areas in their natural state through protected National Forests, National Parks, and National Wildlife Refuges.

The battles between Pinchot and Muir and conservation and preservation of the American landscape led Aldo Leopold to conceive of a more inclusive Land Ethic in his 1939 book *The Sand County Almanac*. Leopold argued that ethics direct people to cooperate for the mutual benefit of all, and that this community, this land, should include all nonhuman elements such as soils, water, and animals. Although widely read and arguably the first environmentalism book, *The Sand County Almanac* failed to mobilize a broad-based coalition to rein in runaway capitalism and industrialization. On the other hand, Rachel Carson's 1962 book *Silent Spring* ultimately galvanized the environmental movement. Carson's book came at a time when many US citizens were interested in and concerned about the environment. The creation of a National Park system in the USA, along with the advent of school and work holidays, and the availability of cheaper automobiles had inspired millions in the post-depression and post-WWII eras to drive out west for yearly vacations. In doing so, the American public became more aware of the plight of the peregrine, bald eagle, buffalo, wolves, and other endangered animals. Carson's book that revealed to the general public and media the catastrophic impacts of DDT was not alone in highlighting the perils of runaway capitalism and industrialization. Apollo 11's photograph of earth rising over the moon's horizon sparked the notion of "spaceship earth" and ultimately contributed to the first

Earth Day. Donella Meadows 1972 book *The Limits to Growth* spelled out in graphic detail several horrific future scenarios for earth given the various resource-consumption projections. All these activities, events, and books, among many others, helped usher in the Clean Air & Water Acts of the early 1970s.

5.9 Ecosystem Services

The value of the earth's support services, such as clean air, clean drinking water, and healthy soil, among a myriad of other services, has been ignored or undervalued for all of human history. But beginning in the late 1990s and early 2000s, their value increased as we realized we were destroying many of their beneficial activities. For instance, deforestation belatedly illustrated the importance of forests to mitigating floods and droughts, reducing soil erosion through rill and gullying and wind erosion, and the silting of rivers and dams. And globally the sequestering of carbon! According to Daily and coauthors, ecosystem services are critical to the continued existence of humanity, and they operate both at a global scale and at a microscopic level such as bacteria detritivores and cannot be replaced by technology. Sadly, human activities have and continue to impair ecosystem services across all scales, and if current trends continue, we will destroy the Earth's surviving natural ecosystems by 2050 if we are to believe the book, *The Limits to Growth*.

According to Cagan Sekercioglu, birds are among the most diverse groups providing ecosystem services. Birds provide bushmeat, feathers for down garments, and even to this day fertilizer in the form of guano. Vultures consume carcasses reducing the spread of rabies. The recent 97% die-off of Indian vultures has cost the Indian government an additional US$34 million in medical bills between 1993 and 2006 due to the surge in rabies across the country as the number of pariah dogs has exploded as cattle carcasses are not being efficiently cleaned by vultures. Insectivores consume invertebrate pests, while many bird species contribute to pollination, especially hummingbirds in the Western Hemisphere and sunbirds in the Eastern Hemisphere. Granivores and seed-eaters, from sparrows and finches to hornbills to oilbirds, consume wild avocados and disperse seeds, while many birds contribute cultural attributes such as feathers to native Hawaiians, native Americans (eagle feathers), and Papuans. Moreover, localities that support bird watching receive millions of dollars in tourism revenue.

▶ Example

Jezeer and coauthors found that five-hectare coffee farms maximize income when at least a third of their land is shade-grown coffee. The shade trees promote healthy soils through higher leaf-fall and subsequent organic decomposition (free of artificial fertilizer), and higher soil moisture through lower evaporation transpiration rates. Birds also contribute by reducing plant and tree pests, in particular, boring beetles and aphids. According to Hernandez-Aguilera and coauthors, a single insectivorous bird can save 10 to 30 kg of coffee per hectare per year. But the birds also significantly reduce or eliminate the use (and cost) of pesticides, which also reduces the cost of labor. So, although coffee yields among shade-grown coffee is typically 30% less than sun-grown coffee, there is a significant reduction in total costs. Shade-grown coffee is typically organic and commands higher prices per kg when sold to European and North American buyers. In contrast, sun-coffee plantations in Central and South American and the Caribbean have 90% fewer birds present and must use pesticides and fertilizers to the tune of $2000 per hectare. With the explosion of sun-grown coffee from Vietnam since the 1980s, many American farmers were forced out of business. Many of those coffee growers in Central and South American and the Caribbean that have survived have turned to organic shade-grown coffee that can also be sold through fair-trade high-end coffee companies. Gustave Axelson interviewed the Fernández family in Costa Rica in 2019 who grow shade-grown coffee. Richard Fernández, the eldest son, noted that they typically have zero coffee beans spoiled by the *broca*, the coffee borer beetle. They get more profits "because the habitat is very comfortable and pleasant for birds." And the family receives two premiums—a Rainforest Alliance certification, and the other from Nespresso who are their main buyer because their farm meets their Nespresso's AAA Sustainable Quality certificate. As one of the benefits, Richard sees toucans nearly every single day. ◄

The ecological services that bats offer frequently mirror those services offered by birds, but bats operate at night. Bats provide bushmeat, although given their colonial nesting and roosting, they provide a fertile setting for virus incubation! In response, bats have fierce immune responses that drive viruses to higher virulence and greater lethality when these viruses infect humans. Nevertheless, bats are significant tropical pollinators, and significant consumers of invertebrate agricultural pests. Interestingly, according to Puig-Montserrat and coauthors, bats have proved especially effective in controlling borer infestations in rice paddies in Spain such that the cost of deploying bat boxes is offset by the reduction in pesticide usage.

The cutting up of our once great forests has left much of our forests as patches or isolates among a matrix of high- and low-intensity agricultural fields and urban sprawl, and as a result, their ecosystem services are much reduced. By the 1980s, conservationists had developed a rule termed SLOSS that stated that a single large reserve is superior to several small reserves of equal size to the large reserve. However, Simberloff found that sometimes a few smaller isolated patches of forest could be more diverse than one big forest patch, and hence the smaller patches were worth protecting. This would depend on the size and distance from an adjacent pool of species in a larger, more intact forest, and the dispersal ability of the species in this larger forest. Furthermore, small patches exhibiting more habitat diversity might offer more habitats than a larger patch. In 1976, Lovejoy set out to address these issues in the Minimum Critical Size Ecosystem Project near Manaus, Brazil. Today, this project is named Biological Dynamics of Forest Fragments Project and it covers some 1000 km^2. Lovejoy's work suggests that one large reserve can hold more species at equilibrium conditions where emigration balances extinctions more than a small reserve. Nevertheless, a reserve located closer to other reserves that are tenuously connected to or at least clustered near each other support more species than a group of reserves that are disjunct or arrayed in a line. A round reserve will hold more species than an elongated one, while a reserve with a river or multiple habitat types will maintain more species than a reserve lacking habitat variation and/or rivers. Lovejoy's work also suggested that a newly isolated reserve will temporarily hold more species than its equilibrium number, but that surplus of species will eventually disappear, as ecosystem decay occurs. The rate at which relaxation occurs will be faster for small reserves than for large ones, while different species require different minimum areas to support an enduring population.

The key element of ecosystem services is the need to understand that very large numbers of species and populations are required to sustain ecosystem services. It is no use just saving giant pandas or elephants; we must save and restore entire ecosystems and not just fragments as Lovejoy's work illustrates. However, minor changes to damaged ecosystems can restore the overall ecosystem to a healthy condition. Yellowstone and the reintroduction of the wolf illustrate the complexities and unexpected changes that can occur even within highly protected forest and wilderness landscapes if these environments are allowed to function as naturally as possible. Although in Yellowstone's case, it is surrounded by a human-mosaic of land uses.

Wolves and Yellowstone

Yellowstone is a microcosm of our modern earth. It is isolated from the surrounding wild matrix of landscapes in Canada and the USA by extensive cattle-utilized rangelands and cultivated and irrigated lands and low-density urban sprawl. And by 1926, the apex predator, the gray wolf, had been eliminated by a government-sponsored eradication program designed to support the adjacent ranchers and protect their cattle. The impact within Yellowstone National Park was catastrophic. Without predatory controls on elk and coyotes, these two animals changed the Yellowstone ecosystem. The elk herd multiplied and changed the spatial nature of forest regeneration within the park, while the growing coyote populations reduced the numbers and diversity of small herbivores and birds throughout the park. Aspens retreated upslope and seed and sapling recruitment ceased due to increasing elk grazing pressures. Elk trampling along and within water courses reduced willow tree density and recruitment, and streams and rivers became muddy and warmer, killing off many native fish. Beaver numbers collapsed as tree recruitment failed around the meadows and valleys; as a result, beaver dams and meadows began to disappear from the park and with them many insect, fish, amphibian and bird species, dependent on these meadows, disappeared as well. Within 70 years of wolf removal, the biodiversity of Yellowstone had crashed, while seed/sapling recruitment among aspen and willows had declined. Even grizzly bears had declined as fewer wolf kills were available to them to scavenge and fewer trees harbored berries in the fall. In essence, a negative trophic cascade had occurred in the 70 years that wolves were absent from the park.

In 1995, this all changed with the reintroduction of wolves into the Yellowstone NP. Although wolves are not numerous throughout the park, their impact and resulting positive trophic cascades has begun to naturally repair a damaged ecosystem. The wolves kill the young, weak, and old mule deer, elk, and moose. These deer now avoid the valleys and gorges where they might be trapped by hunting wolves, and as a result, aspen, cottonwood, and willow trees have regenerated; tree height has grown fivefold since 1995. Beavers have returned and now forage on the new sapling and growing trees. The new beaver dams have become havens for otters, muskrats, ducks, and fish. The dams have improved water quality by trapping sediment, replenishing groundwater, and cooling water. Trout and other fish have returned to the park.

Wolves have also predated on coyotes, and this has led to more rabbits and mice, which led to more hawks, weasels, badgers, and foxes. The wolves have left more carrion and that has benefited both bears and bald eagles. The bears also have more access to fruit and berries resources during their feeding frenzy before hibernation. Finally, and most unexpectedly, the wolves have altered stream and river dynamics because they have excluded deer from the valleys. Rivers braid less, suffer less bank erosion, and exhibit deeper pools and more riffles as regenerating trees have strengthened stream and river banks.

Hunting guides and their clients have benefited in two ways. First, the wolves have helped to create a healthier population of mule deer, elk, and moose by removing the less desired animals from the gene pool, and second, the hunting guides now also guide tourists to see and photograph (but not kill) wolves.

5

Conclusion

In summary, climate change is likely to lead to major populational health impacts, at the same time that we are at a peak energy crisis (peak oil production occurred in 2007) and we have exhausted many of the natural resources that allowed us to achieve our current state of civilization. We are also rapidly approaching the Earth's carrying capacity and capacity for ecological services to save us. However, recent reductions in GHG emissions and coal-generated electrical production, increases in renewable energy forms such that several countries routinely produce 100% of their electrical needs from sustainable energy forms, and wolf introduction to Yellowstone hold out hope that at the very cliff of environment destruction, we might be able to save ourselves and others!

But we are failing to save our tropical lowland forests, our last great site for terrestrial carbon sequestration. We cannot allow business-as-usual approaches to resource consumption to continue, we cannot let *The Limits to Growth* accurately predict the demise of our civilization and the rampant and deadly pollution of the earth by our industrial and consumptive activities.

References

Ambrose, J., & Goodley, S. (2020). Carbon emissions fall as electricity producers move away from coal. *The Guardian*, 9 Mar 2020. https://www.theguardian.com/environment/2020/mar/09/carbon-emissions-fall-as-electricity-producers-move-away-from-coal?CMP=Share_iOSApp_Other

Amelang, S. (2018). Renewables cover about 100% of German power use for first time ever|. *Clean Energy Wire*. https://www.cleanenergywire.org/news/renewables-cover-about-100-german-power-use-first-time-ever. Accessed 23 Jan 2019.

Axelson, G. (2019). Coffee made in the shade can be more profitable, thanks to birds. Living Bird magazine. https://www.allaboutbirds.org/news/coffee-made-in-the-shade-can-be-more-profitable-thanks-to-birds/

Bar-On, Y. M., Phillips, R., & Milo, R. (2018). The biomass distribution on earth. *Proceedings of the National Academy of Sciences*, 201711842.

Bosak, K. (2004). Biodiversity conservation and the struggle for the Nanda Devi biosphere reserve. *Focus on Geography, 48*(1), 1–6.

Bremmer, I. (2006). *The J curve: A new way to understand why nations rise and fall*. Simon & Schuster.

Carrington, D. (2018). Humans just 0.01% of all life but have destroyed 83% of wild mammals – Study. *The Guardian*.. https://www.theguardian.com/environment/2018/may/21/human-race-just-001-of-all-life-but-has-destroyed-over-80-of-wild-mammals-study. Accessed Sept 26, 2018.

Collins, M. (2005). Life cycle assessment of disposable and reusable nappies in the UK. *Environment Agency*.

Crump, A., & Omura, S. (2011). Ivermectin, 'wonder drug' from Japan: The human use perspective. *Proc Japan Academy, Series B, 87*(2), 13–28.

Cutter, S. L., & Renwick, W. H. (1999). *Exploitation, conservation, preservation: a geographic perspective on natural resource use (No. Ed. 3)*. John Wiley and Sons.

Daily, G. (2003). What are ecosystem services. *Global environmental challenges for the twenty-first century: Resources, consumption and sustainable solutions*, pp. 227–231.

Enloe, C. 2001. *Bananas, beaches and bases: Making feminist sense of international politics*. 263 p. University of California Press.

EPA. (2016). *Climate change indicators: U.S. greenhouse gas emissions*. https://www.epa.gov/climate-indicators/climate-change-indicators-us-greenhouse-gas-emissions. Accessed 23 Jan 2018.

Foale, S., & Macintyre, M. (2005). Green fantasies: Photographic representations of biodiversity and ecotourism in the Western Pacific. *J Politic Ecol, 12*, 1–22.

Gore, A. (2006). *Am inconvenient truth: The planetary emergency of global warming and what can we do about it*. Rodale.

Gubler, D. J. (1998). Resurgent vector-borne diseases as a global health problem. EID vol 4(3): Special Issue.

Gunders, D. (2012). Wasted: How America is losing up to 40 percent of its food from farm to fork to landfill. *Nat Res Def Council, 26*.

Heinberg, R. (2003). *The party's over: Oil, war and the fate of industrial societies*. NSP.

Hernandez-Aguilera, J. N., Conrad, J. M., Gómez, M. I., & Rodewald, A. D. (2019). The economics and ecology of shade-grown coffee: A model to incentivize shade and bird conservation. *Ecological Economics, 159*, 110–121.

Honey, M. (1998). *Ecotourism and sustainable development: Who owns paradise?* Island Press. 416 p.

Hu, Z.-Z., Yang, S., & Wu, R. (2003). Long-term climate variations in China and global warming signals. *Journal of Geophysical Research, 108*(D19), 4614. https://doi.org/10.1029/2003JD003651.

IHA. (2019). *Hydropower status report: Sector trends and insights*. https://www.hydropower.org/sites/default/files/publications-docs/2019_hydropower_status_report.pdf

Isaacs, J. C. (2000). The limited potential of ecotourism to contribute to wildlife conservation. *The Ecologist, 28*(1), 61–69.

Jones, D., Huscher, J., Myllyvirta, L., Gierens, R., Flisowska, J., Gutmann, K., Urbaniak, D., & Azau, S. (2016). *Europe's Dark Cloud–how coal burning countries are making their neighbours sick*.

Kamauro, O. (1996). *Ecotourism: Suicide or development? Voices from Africa #6: sustainable development, UN non-governmental liaison service*. United Nations News Service.

Klein, N. (2007). *The shock doctrine: The rise of disaster capitalism*. Metropolitan books.

Knack, S., & Keefer, P. (1997). Does social capital have an economic payoff? A cross-country investigation. *The Quarterly Journal of Economics, 112*(4), 1251–1288.

Kunstler, J. H. (2005). *The long emergency: Surviving the end of oil, climate change, and other converging catastrophes of the twenty-first century*. NY: Grove Press.

Lazaridis, I., Nadel, D., Rollefson, G., Merrett, D. C., Rohland, N., Mallick, S., Fernandes, D., Novak, M., Gamarra, B., Sirak, K., & Connell, S. (2016). Genomic insights into the origin of farming in the ancient near east. *Nature, 536*(7617), 419.

Lehrburger, C., Mullen, J., & Jones, C. V. (1991). *Diapers: Environmental impacts and lifecycle analysis (no. 677.21 L524d). Massachusetts, US: Sn* (p. 1991).

Mann, M. E., & Kump, L. R. (2009). *Dire predictions: Understanding global warming*. IPCC.

Maplecroft, V., 2011. Climate change vulnerability index 2016. *Climate Change and Environmental Risk Atlas*.

Micklin, P. (2007). The Aral Sea disaster. *Annual Review of Earth and Planetary Sciences, 35*, 47–72.

Mirovalev, M. (2015). *Uzbekistan: A dying sea, mafia rule, and toxic fish*. AJImpact. https://www.aljazeera.com/indepth/features/2015/06/uzbekistan-dying-sea-mafia-rule-toxic-fish-150610102819386.html.

Mullner, A., Linsenmair, K. E., & Wikelski, M. (2004). Exposure to ecotourism reduces survival and affects stress response in hoatzin chicks (Opisthocomus hoazin). *Biological Conservation, 118*, 549–558.

New, M., Liverman, D., Schroder, H., & Anderson, K. (2011). Four degrees and beyond: the potential for a global temperature increase of four degrees and its implications. *Phil. Trans. R. Soc. A, 369*, 6–19.

NSIDC Notes (2010) Issue 73. nsidc.org

Pratt, K. (2006). Tourism to altruism: The emergence of private protected areas in Chile. AAG 2006 Abstracts Online. http://www.aag.org/

Puig-Montserrat, X., Torre, I., López-Baucells, A., Guerrieri, E., Monti, M. M., Ràfols-García, R., Ferrer, X., Gisbert, D., & Flaquer, C. (2015). Pest control service provided by bats in Mediterranean rice paddies: Linking agroecosystems structure to ecological functions. *Mammalian Biology, 80*(3), 237–245.

Ranganathan, J., Vennard, D., Waite, R. I. C. H. A. R. D., Dumas, P., Lipinski, B., & Searchinger, T. (2016). *Shifting diets for a sustainable food future*. World Resources Institute.

5

Rauch, P. S. (2016). *Forsaken earth: The ongoing mass extinction*. Lulu.com.

Sakata, H., & Prideaux, B. (2013). Community-based ecotourism: Opportunities and difficulties for local communities and link to conservation. In B. Prideaux (Ed.), *Rainforest tourism, conservation and management: Challenges for sustainable development* (pp. 199–212). Routledge: The Earthscan Forest Library.

Salcito, K. 2015. Why cloth diapers might not be the greener choice, after all. Washington Post, May 8, 2015. Accessed 17 Oct 2018.

Saha, N., Mollah, M. Z. I., Alam, M. F., & Rahman, M. S. (2016). Seasonal investigation of heavy metals in marine fishes captured from the bay of Bengal and the implications for human health risk assessment. *Food Control, 70*, 110–118.

Sarraf, M., Stuer-Lauridsen, F., Dyoulgerov, M., Bloch, R., Wingfield, S., & Watkinson, R. (2010). *The ship breaking and recycling industry in Bangladesh and Pakistan*. Washington, DC. World Bank: Washington, DC

Sekercioglu, C. H. (2002). Impacts of birdwatching on human and avian communities. *Environmental Conservation, 29*(3), 282–289.

Sekercioglu, C. H. (2006). Increasing awareness of avian ecological function. *Trends in Ecology & Evolution, 21*(8), 464–471.

Sippel, S., Meinshausen, N., Fischer, E. M., Székely, E., & Knutti, R. (2020). Climate change now detectable from any single day of weather at global scale. *Nature Climate Change, 10*(1), 35–41.

Sovacool, B. K. (2008). Valuing the greenhouse gas emissions from nuclear power: A critical survey. *Energy Policy, 36*(8), 2950–2963.

Stiglitz, J. E. (2002). *Globalization and its discontents*. New York: W.W. Norton.

Meadow, D., Randers, J., & Meadows, D. (2004). *Limits to growth: The 30-year update*. White River Junction: Chelsea Green.

Sutherst, R. W. (2004). Global change and human vulnerability to vector-borne diseases. *Clinical Microbiology Reviews, 17*(1), 136–173.

Theil, S. (2008). Germany: Best Governed Country In Environment. Newsweek. https://www.newsweek.com/germany-best-governed-country-environment-91161. Accessed 23 Jan 2019.

Tisdell, C. (2003). Economic aspects of ecotourism: Wildlife-based tourism and its contributions to nature. *Sri Lankan Journal of Agricultural Economics, 5*(1), 83–95.

Vivanco, L. (2002). Ecotourism, paradise lost - a Thai case study. *The Ecologist, 32*(2), 28–30.

Vorrath, S. (2018). Portugal reaches 100% renewables, ends fossil fuel subsides. Renew Economy 9 April 2018. https://reneweconomy.com.au/portugal-reaches-100-renewables-ends-fossil-fuel-subsidies-32820/

Walker, B. G., Boersma, P. D., & Wingfield, J. C. (2005). Physiological and behavioral differences in magellanic penguin chicks in undisturbed and tourist-visited locations of a Colony. *Conservation Biology, 19*(5), 1571–1577.

Wallerstein, I. (1987). *World-systems analysis*. Duke University Press.

Webster, P. J., Holland, G. J., Curry, J. A., & Chang, H.-R. (2005). Changes in tropical cyclone number, duration, and intensity in a warming environment. *Science, 309*(5742), 1844–1846.

WWAP, U. (2015). The United Nations world water development report 2015: Water for a sustainable world. *United Nations World Water Assessment Programme*.

Ziffer, K. (1989). *Ecotourism: The uneasy alliance*. Conservation International/Ernst and Young.

Population

Contents

M. R. Welford, R. A. Yarbrough, *Human-Environment Interactions*,
https://doi.org/10.1007/978-3-030-56032-4_6

Learning Goals

After reading this chapter, you will be able to:

- Explain the history of global population growth.
- Explain the demographic transition theory (and its critiques).
- Examine contemporary geographic patterns of population and population growth.
- Evaluate multiple perspectives on population, resources, and the environment.
- Illustrate the IPAT model and apply it to today's world.

Comprehension of population patterns and processes is integral to any discussion of human-environment interactions, whether global in scope or at regional, urban, or more localized geographic scales. Indeed, with nearly 7.5 billion inhabitants of the earth, numbers remain a significant piece of the human-environment puzzle. Nevertheless, numbers alone often belie the complexities and contingencies of human-environment interactions. The following discussion of geographic patterns of fertility, mortality, and migration, therefore, provides a foundation for understanding how, why, and where people affect and are affected by environmental processes at multiple geographic scales.

6.1 Contemporary Geographies of Population

At the time of this writing (late 2019), our planet had over 7.5 billion inhabitants, distributed quite unevenly amidst the world's geographic regions. Indeed, nearly six in ten of the earth's residents live on one continent, Asia, with China accounting for 19%, India 18%, and the rest of Asia 21%. The entirety of Africa accounts for 17% of the world's population, while North America, Oceania, and Europe combine for only 16%. That leaves 9% of the earth's inhabitants residing in Latin America (US and World Population Clock).

As you may know, both China and India are home to well over a billion persons today, yet China will soon be surpassed by India (likely by 2025) as the world's most populous country. By then, if not shortly thereafter, China is expected to reach a stable population with no annual growth, also known as zero population growth (ZPG) and may even begin declining by the year 2030. The most reliable projections of the future of the global population expect that we will reach (or come very close to) 10 billion by the year 2050, with much of that growth occurring in Africa.

Indeed, over the next 30 years, Africa is expected to account for approximately 60% of the world's total population growth. By contrast, Europe is the only region expected to have a smaller population in 2050 than it has today. Why is Europe's population going to shrink? Simply put, they are not having enough children to maintain a stable population. The good news for Europe? People are living longer and longer. We see a similar trend throughout most of the world as life expectancies continue to increase. For instance, the world's 80+ population is likely to grow threefold by 2050. Combine that with people having fewer and fewer children (i.e., total fertility rates are declining in all regions of the world and that trend is likely to continue), and we see a demographic situation in which the world is getting older. Specifically, the median age will go from 30 years old in 2018 to 37 by 2050. So, although only 13% of today's world residents are aged 60 or older, these older persons will account for over ¼ of the global population in 2050 (United Nations Department of Economic and Social Affairs, Population Division 2017).

To summarize, the world's population will continue to grow throughout the twenty-first century, but the pace of that growth will be slower than it is today. Although Europe will actually experience a declining population, most regions will continue to add residents, and the earth will have to cope with a much larger population, likely cresting at 11 billion by the beginning of the next century. But how did we get to this point and what are the implications of the demographic situation we find ourselves in at this time in human history? It is to these questions that we now turn for the remainder of the chapter.

6.2 A Brief History of Global Population Growth

Modern humans (*Homo sapiens*) have inhabited the earth for at least 200,000 years, and the primary components of population change, fertility, mortality, and migration have interacted in a particular relationship with one another for the vast majority of that time. First, we should note the obvious—migration is irrelevant when discussing global population growth (i.e., no one is migrating to or from the earth—at least *not yet*). So that leaves fertility and mortality, birth rates and death rates, as the only components of demography that have affected the number of people living on the planet throughout history. For most of human prehistory (i.e., before the earliest known forms of writing emerged around 3000 BCE that enabled the recording of history), fertility was relatively high, influenced primarily by biological factors (what demographers call fecundity or the physical ability to reproduce) and frequency of sexual intercourse; these were the main drivers of birth rates. Perhaps not surprisingly, fertility rates were quite high, since there was very little effort to plan the number of children women had. What may surprise you, however, is that these high birth rates led to very little notable growth in the world's population. Why? Because effectively they were offset by very high mortality rates.

> The mortality rate (also known as death rate) is defined as the number of deaths per 1000 population in a given year.

Think for a moment about what life must have been like for the earliest humans on earth—to say it was a tough life would be a massive understatement. Food supplies were not always abundant, there was no real knowledge of disease or modern medicine, and life was quite violent in many cases. These very difficult living conditions translated into high infant mortality rates (the number of babies dying before the age of 1, per 1000 live births) and very low life expectancies. In short, life was very hard for most, and also very short. So, despite very high birth rates throughout most of human history, global population growth remained relatively low because these were cancelled out by high mortality rates. However, the relationship between birth rates and death rates began to change dramatically around 10,000 BCE with the coming of the First Agricultural Revolution.

The First Agricultural Revolution. Beginning approximately 12,000 years ago, the First Agricultural Revolution (also known as the Neolithic Revolution), was characterized by tremendous demographic, social, and environmental changes and brought about a major increase in the world's population. While in 10,000 BCE the world's population was only about 5 million people, significant advancements in the domestication of plants and animals began to occur in multiple geographic locations that would lead to the emergence of settled civilizations.

> Anthropologists and historians generally cite the First Agricultural Revolution as the period marking an overall shift from hunter-gatherer societies to sedentary, stable settlements, which spurred global population growth because of the newfound ability to produce surplus amounts of food and thus support larger and larger populations ("The Genographic Project" National Geographic).

Along with these agricultural and demographic transformations came social and economic changes in the form of divisions of labor across societies (i.e., because of the shift to stationary settlements, specific social structures developed around certain people doing certain jobs within a society). By the time the Medieval Black Death struck in the 1300s, the world's population had grown to an estimated 450 million, but by the time it had subsided, the number of living persons on earth had declined by nearly one-fourth.

■ *The Black Death and Population Decline*

The Medieval Black Death (MBD), the plague that first hit Europe in 1347 and returned episodically and lethally until 1815, killed some 75 million people in Europe (Scott and Duncan 2004). From the initial primary wave of the MBD that rampaged through Europe from 1347 to 1353, Europe did not recover demographically until the mid-1500s. In fact, MBD hitched a ride from Central Asia and its endemic heartland to Europe along both the overland and maritime Silk Network. In many ways, MBD was an unforeseen and deadly consequence of proto-globalization as the Old World began to transition from a world with limited connections to a world that was becoming increasingly spatially integrated. The Silk Network was the "super-highway" of its time, providing a conduit for trade, information, technology, people, and disease to move rapidly back-and-forth across Eurasia.

Although the MBD killed millions, the scale and deadliness of plague varied spatially across Europe—geography very much mattered. Britain, particularly East Anglia, was especially hard hit, suffering MBD mortalities in excess of 50%, in part due to its rich agricultural landscape that facilitated a high density of markets. Indeed, big cities, especially ports and capitals, suffered longer epidemics than smaller, less well-connected cities. For instance, those towns or cities on medieval pilgrimage routes suffered very high mortalities, yet towns, villages, or hamlets adjacent to such routes but not frequented by religious travelers escaped with relatively few deaths. The early implementation of quarantine by Pistoia and Milan in 1348 saved these towns from significant mortality. Majorca, Venice, and others quickly followed their example: travelers suspected of carrying plague were stopped from entering those cities, while residents of those cities displaying plague symptoms were isolated in their houses. These tactics proved very successful in preventing the outbreak of MBD. By the 1600s, many European nations adjacent to the Balkans periodically implemented aggressive disease cordons to isolate themselves when plague was identified by their plague spies.

The MBD was not an isolated event—it happened during a period of intense climatic instability. First, the accelerated melting of the Greenland icecap due to the Medieval Warm Period shifted the Gulf Stream south as cool, less dense, lower salinity surface waters expanded across the northern Atlantic. This increased climatic variability led to cooler temperatures and reduced precipitation across western and central Europe, one consequence of which was the Great Famine of 1315-1318. Second, the 1258 eruption of an Indonesian volcano cooled the entire globe by several degrees. This period is known today as the Little Ice Age and ran from 1300-1800. As a result, agricultural output fluctuated wildly throughout the Little Ice Age, causing periodic and lasting food shortages that led most peasants to suffer compromised immune systems that were frequently unable to survive plague infections (Welford 2018).

Columbian Encounter (Near Extinction). You may have heard the phrase "In 1492 Columbus sailed the ocean blue." This is a mnemonic device taught to many elementary school children in the USA to help remember when Christopher Columbus landed in the current-day Bahamas. This so-called discovery of the New World marked the beginning of a series of significant and long-lasting interactions between Europe and the Americas, including colonization and the trans-Atlantic slave trade. With Columbus came a near extinction of the indigenous inhabitants of the Americas in a matter of decades due to their exposure to Old World pathogens (that they had no immunity to) such as measles, flu, and smallpox. Researchers have estimated a decrease in the indigenous population of the Americas from approximately 55 million at the time of Columbus' arrival to 5-6 million by 1600—a decline of 90% (Lovell 1992)! This massive demographic collapse resulting from the Columbian exchange represented about one-fifth of the world's population at the time. Not only was this a horrific and substantial demographic event, it also resulted in monumental cultural losses. Indigenous knowledges, ways of life, and cultural beliefs and practices were lost along with those lives or were altered considerably with the coming of the colonial settlers.

Tragically, this enormous loss of life was followed by another, as the colonial settlers turned to Africans as a means to replace this lost labor supply.

According to Koch et. al. (2019), the great dying in the Americas after 1492 also contributed to the Mini Ice Age as 55.8 million hectares of pre-Colombian agricultural land reforested, leading to the uptake and sequestering of 7.4 Pg (or 7.4 billion tons) of carbon.

Finally, the Columbian exchange was a precursor to today's increasingly globalized world and the near insurmountable challenges that have emerged from unfettered economic growth, with very little by way of global concentrated efforts at ameliorating the resulting environmental damage.

The Industrial Revolution, The First One billion, and Today

A second Agricultural Revolution, which saw increased yields primarily through greater mechanization, occurred around the time of the Industrial Revolution in the late eighteenth and early nineteenth centuries. Yet it was not this second revolution in agriculture that led to notable population growth, but rather overall improvements in standards of living resulting from the Industrial Revolution that led to major increases in life expectancy and therefore decreases in death rates. Indeed, the total number of people on earth reached one billion by 1800, then continued to grow exponentially by the early twentieth century, due primarily to declines in mortality (people living longer) from advancements in medicine (e.g., knowledge of germs). The next few billion persons were added relatively quickly compared to the previous trend; the 2 billion mark was reached in 1930, 3 billion in 1960, and by the twenty-first century, the world's population had surpassed 6 billion.

The annual growth rate of our global population peaked in the late 1960s at 2.06% and has been falling steadily since then. However, as it remains above zero, we continue to add to the global population at a rate of just over 1% annually (1.09% in 2018). This trend highlights the difference between declining population growth and a declining population.

> When the pace at which a population is increasing declines over time, this represents decreasing population growth (as in the case just mentioned—going from 2.1% annual global population growth in 1961 to 1.09% annual growth in 2018). Of course, we are still adding people to the earth's population on a daily basis; in fact, it is approximately 215,000 each day! Hence, the world's population of approximately 7.456 billion in 2018 will continue to grow throughout the twenty-first century. For examples of declining populations, we must look to several individual countries like Greece, Hungary, Estonia, and Japan, where the number of residents has been decreasing in recent years because of a combination of low fertility, high mortality (primarily because of the number of elderly residents) and/or high out-migration.

In conclusion, the three major moments in global population growth were Neolithic Revolution (First Agricultural Revolution) approximately 10,000 years ago, the Industrial Revolution in the late 1700s and early 1800s, and the early to mid-twentieth-century medical and scientific advances (knowledge of germs, etc.). Only the First Agricultural Revolution impacted birth rates to spur population growth, while world population growth in the latter two moments was brought about by increasing life expectancies resulting in a decline in mortality rates.

6.2.1 The Demographic Transition

The Demographic Transition (DT) model explains how societies move from a state of high birth rates and high death (mortality) rates to a state of both low birth rates and death rates. The DT explains that these demographic changes occur with increases in economic development, particularly industrialization, increased urbanization, and changes to cultural norms associated with family size.

Traditionally, the DT was composed of four phases, although more recently many have added a fifth phase to represent a large portion of the developed world today that is experiencing negative population growth (i.e., declining populations). The DT model presumes that countries move (or will move) from one phase to the next until they reach phase five. But as we will see, there exists a great deal of geographic and temporal variation when we examine the demographic stories of individual countries.

Demographic Transition

The main characteristics of the five phases of the Demographic Transition are:

Phase 1—High birth rates and high death rates lead to very little or zero population growth (ZPG).

Phase 2—Importantly, death rates begin to fall rapidly, while birth rates remain relatively stable. This precipitous decline in mortality rates leads to a high population growth rate (i.e., when fewer people are dying in any given year, but the same number of babies are being born, the result is notable population growth).

Phase 3—Mortality rates begin to level off and birth rates begin a steady decline. The result is an increase in population, but at a *slower rate* than Phase 3. Effectively, birth rates are declining to catch up to death rates, which leads to a lower growth rate than exhibited in Phase 2.

Phase 4—Birth rates have declined to an equilibrium with mortality rates, thus leading to very low or no population growth from natural increase.

Phase 5—Mortality rates remain low and stable, while birth rates may decline slightly. Over time, this potentially results in negative population growth (declining populations).

In today's world, no countries remain in Phase 1 of the DT, although most in the Global North moved into Phase 2 and beyond much earlier than did countries in the Global South. Indeed, today most developing countries are either in Phase 2 or 3, while some have moved into Phase 4. By contrast, most Global North countries exhibit demographic characteristics described in Phase 4 or 5 according to the DT model. It is also important to note that many of the factors affecting birth rates and mortality rates may not necessarily be associated with industrialization and economic development. For example, regardless of changes to the economic structure and standard of living in a society, there exist myriad social, economic, and cultural factors that influence family size decisions (i.e., birth rates). Indeed, the influence of family planning programs, women's education, and gender roles (both within the family and across society more broadly) have been found to indirectly impact direct determinants of fertility such as contraceptive use and frequency of intercourse.

Several caveats apply when discussing how applicable the DT is to the actual demographic experiences of countries in today's world. First, it is crucial to note that the model itself is based upon the experiences of developed countries, especially the United States and Western European nations. Thus, it should not come as a surprise that the DT may not perfectly map onto the demographic histories and experiences of many countries in the developing world. Indeed, many countries in the Global South are experiencing some type of fertility transition (in general, birthrates are declining worldwide) in the absence of large-scale economic development and industrialization. This is the case in countries as diverse as Guatemala and Kenya, wherein significant fertility declines have brought total fertility rates (the average number of children born) down to below 3 in the absence of significant widespread industrialization or economic development. Some counterexamples exist as well in relatively wealthy, industrialized countries that continue to exhibit high fertility rates notwithstanding significant industrialization and economic development (e.g., Qatar).

Despite these legitimate critiques of the Demographic Transition, the model remains useful in comparing countries' population patterns and helping to predict their future demographic changes. Moreover, it can serve as a point of departure for discussing the complex relationship between demographic trends and environmental impacts at multiple geographic scales (i.e., from the global to the local). Initially, it may be tempting to conclude that these are almost always positively correlated; that is, that if population increases, so does environmental impact and vice versa. To borrow a term from the discipline of economics, "all things being equal" (from the Latin ceteris paribus), a change in population would likely lead to an associated change in environmental impact. However, drawing such a conclusion in a real-world context oftentimes results in a misleading and overly simplistic view of the role of population in environmental degradation. Indeed, as we have noted in previous chapters and elaborate on later in the text, we must also critically examine consumption practices, waste disposal, and technology, among other crucial factors, in this complex relationship between population and the environment.

6.3 Perspectives on Population, Resources, and the Environment

Thomas Robert Malthus

Strange as it may seem, any discussion of population and its relationship to resources must begin with an introduction to the writings of a minister in England who lived in the late eighteenth and early nineteenth centuries. Thomas Robert Malthus (he actually went by his middle name, Robert) was a clergyman and professor (it was common for many college professors to be members of the clergy at the time) who published his first essay anonymously in 1798. Because its title "An Essay on the Principle of Population as It Affects the Future Improvement of Society; with Remarks on the Speculations of Mr. Godwin, M. Condorcet, and Other Writers" is quite long and cumbersome, it is usually referred to simply as "An Essay on the Principle of Population." Some historical context is helpful here. In the late eighteenth century, seminal Enlightenment thinkers, including David Hume and Jean Jacque Rousseau, had recently passed away, while emerging figures like William Godwin and Marquis de Condorcet were prominent philosophers at the time. Godwin and Condorcet both argued that reason and intellectual pursuits would bring about great social progress (especially in Europe), a perspective which flew in the face of the harsh living conditions that most faced in Europe at the time. Indeed, by the early nineteenth century, a series of "Poor Laws" were being debated in England as a means of alleviating the widespread ills of persistent poverty amongst many residents. What does all this have to do with population? Well, quite a lot as it turns out. Essentially, Malthus believed that poverty was primarily a function of overpopulation and that any law intended to provide for the most impoverished in society (i.e., "Poor Laws") was simply making matters worse by removing an incentive to have fewer children. In his "Essay on Population" and subsequent writings that quickly followed, Malthus laid out what he argued was a fundamental (and intransigent) relationship between population growth and the food supply, one that has served as the foundation for nearly all discussions of the population-resource conundrum over the past two centuries.

Malthus postulated that while the food supply grows at a linear (or arithmetic) rate, populations grow at an exponential rate (geometrically). This difference in growth rates leads to an unavoidable conclusion—eventually, the population will outgrow the food supply, resulting in a deficit of food. Thus, Malthus' principle on population growth, if left unchecked, necessarily led to a series of fairly depressing conclusions with regard to human suffering (e.g., famines, disease, and wars over food shortages). Ironically, these negative results are examples of what Malthus described as positive checks. To be clear, Malthus was not suggesting that these were positive in a normative sense; he deemed them positive because they resulted in an increase in mortality rates (i.e., they had a mathematically positive impact on the death rate).

He contrasted these positive checks on population growth with preventive checks, which included abstinence (what Malthus referred to as "moral restraint"), abortion, and delaying marriage.

As a minister in the Church of England, Malthus did not believe in abortion and had little faith that the masses could practice "moral restraint," which meant that delaying marriage was the only preventative check that was likely to be implemented.

Neo-Malthusian Perspective. Fast-forward to the 1960s and Malthus' population-resource principle was given a new life with a slight twist. Amidst concerns of global population growth fueled primarily by developing countries that had experienced a decline in mortality, but not yet a notable fertility decline (recall the Demographic Transition model), scientists and activists in developed countries like the USA began sounding the alarm. This group believed wholeheartedly in Malthus' population-resource principle, although they expanded Malthus' focus on the food supply to encompass the use of resources more broadly and the carrying capacity of the earth. These devotees of Malthus were deemed Neo-Malthusians and differed from Malthus in one fundamental and very significant way—they believed in modern-day birth control methods.

Indeed, some went so far as to argue that if families in developing countries did not begin to more fully incorporate modern-day contraceptive practices in order to bring about a fertility transition (decline), it might be necessary to compel them to do so. As extreme as it may sound today (and it did to many at the time), the Neo-Malthusians saw global population growth as a life or death matter for the planet. They were absolutely convinced that without large-scale action to decrease global birth rates, the carrying capacity of the earth would be surpassed and doom was imminent.

Scholars like ecologist Garret Hardin and biologist Paul Ehrlich represented the Neo-Malthusian position, publishing seminal works like "The Tragedy of the Commons" (1968) and *The Population Bomb* (1968), respectively (in fact, Ehrlich's wife Anne Ehrlich actually co-wrote *The Population Bomb*, but was not credited with co-authorship). Hardin's "Tragedy of the Commons" was published in the journal *Science* in late 1968 and received considerable attention both within and outside of academia. Hardin's central premise was that if a communally used resource was left unregulated, individuals would use it to meet their own needs with little to no regard for its sustainability. If all individuals acted as such, it would inevitably be depleted or destroyed. Using the example of a common-use pasture and cattle, Hardin surmised that each herdsman would continue to add cattle to his herd as long as he believed it was in his (individual) best interest. Acting in such a way, however, would continue as the benefit of adding one cow to his herd would go only to him, while the cost of overgrazing would be shared by all who used the commons. Eventually, Hardin (1968) concluded, the cumulative effect of each herdsman acting rationally in his own best interest is the destruction of the common-use field, due to overgrazing. As Hardin (1968) famously stated, "Freedom in a commons brings ruin to all." In this case, Hardin's fundamental argument was that "the freedom to breed" was the freedom that would bring ruin to us all by leading to overuse of the earth's resources (and our ability to dispose of all the waste from consumption of said resources). Thus, his parable of the overuse of the commons served as a rallying cry for Neo-Malthusians who latched onto Hardin's (1968) recommendation of "mutual coercion, mutually agreed upon" in the context of what the United Nations had identified all the way back then: a human right to reproduce.

These leaders of the Neo-Malthusian movement and their doomsday scenarios including potential famines, water shortages, plagues, and/or wars over resources gained widespread attention over the next few decades. Indeed, many of the same arguments regarding population growth and its myriad environmental impacts served as motivation for the first Earth Day in 1970. Despite these doomsday scenarios not coming to fruition on a global scale, this view is still advocated by many and population growth is still targeted as a fundamental global problem. Although unknown at the time, the global population growth rate would peak in 1968 at 2.06% annual growth, and as you now know, that rate would continue to decrease steadily to its current rate of just over 1%. An alternate perspective on the population-resource relationship focuses much less on the number of people and looks much more closely at the uneven consumption patterns and environmental impacts across the globe.

Marxist/Distributionist Perspective. Malthus' premise gained so much attention and became so prominent that myriad intellectuals commented on it throughout the nineteenth century. Despite the fact that population was not central to most of his writings, even Karl Marx commented on Malthus' work, primarily because of the latter's arguments regarding social safety net programs (e.g., the Poor Laws) incentivizing couples to have children. Marx's critiques of Malthus, however, hold credence still today and provide the foundation for a Marxist (or Distributionist) perspective on the population-resource relationship. This Distributionist perspective does not accept the basic Malthusian premise that population will always outgrow the resource base; indeed, it does not really focus on population per se. Rather, a Distributionist perspective critically analyzes the (geographically) uneven consumption of resources and, by extension, the uneven geographies of environmental degradation. For a Distributionist, the question is not *how many* people, but *which* people and *what* are they doing that lead to the consumption of resources and by-products that impact the environment.

Take for instance the West African country of Mali. In 2017, Mali had one of the highest fertility rates in the world, with the average woman having approximately 6 children (i.e., the total fertility rate is 6.01 in 2017). Yet, when we shift the focus to the consumption of resources, be it food, water, or electricity, and consider the by-products of those consumption patterns, the average child in the United States is very likely to have 5-10 times the impact on the earth than the average child in Mali. To take one simple way of measuring this impact, carbon footprint, there is no doubt that the two young daughters of one of the authors of this textbook have a much greater negative impact on the earth than the six children of the average Malian woman. Therefore, it is both misleading and unfair to focus primarily on the numbers—in this case, six children in Mali and two in the United States—if we truly want to analyze the environmental impact of population. From the perspective of many geographers, a Distributionist framework provides the tools to examine the uneven geographic patterns of population, consumption, and environmental degradation in a way that accurately reflects the simple reality of the world in which we all live—that different places contribute more to environmental problems than others. Demographically, this often entails a consideration of per capita (i.e., per person) consumption patterns alongside any analysis of population. This Distributionist perspective does not advocate any particular solution; rather, it asks us to take a critical look at who is doing what in the world and how those impacts (oftentimes of relatively small proportion of the whole) are disproportionately contributing to the world's environmental problems.

Economic Optimists. In sharp contrast from Neo-Malthusians and Distributionists, a third camp of Economic Optimists (sometimes called Technocrats) do not view either population or consumption as problematic in and of themselves. This view questions the validity of the fundamental Malthusian premise and takes as evidence the technological advances in agriculture and resource use that Malthus never anticipated. Thus the Economic Optimist argues that Malthus was simply incorrect in his original premise because resources are not limited to only linear growth—indeed, they have grown exponentially in many cases (e.g., global food production). Moreover, Economic Optimists put great faith in what they see as the most valuable resource of all, the human brain.

Advocates of this perspective, such as the late economist Julian Simon, point to the discovery of new resources, the development of new efficiencies and technologies, and the power of human ingenuity to solve whatever population-resource conundrum comes our way as a human race.

One danger of putting all your eggs in the Economic Optimist basket, so to speak, is that it can lead to very little environmental awareness or change in behavior because it presumes that humans will always find a technological fix. With technology's ubiquity in today's world, this perspective is arguably even more appealing (and somewhat concerning) than it was during Simon's heyday of the 1980s. Certainly, this is the case with climate change—scientists doubt that there is some yet-to-be-discovered technology that will become the magic wand we can waive and solve the impending climate crisis. Yet, in many developed countries, including the United States, there is very little evidence that people are changing their behavior relative to its environmental impact. No doubt some of this is related to domestic politics, but much of it is simply ignorance, indifference, and/or selfishness. As with most human behaviors, change only comes when it either becomes less beneficial or even harmful to continue acting in the same way. The Economic Optimist perspective holds the potential to keep us from forcing ourselves to change our behavior until it is too late, because of our (likely misguided) faith that technology will save us.

It is important to note that there are components of all three perspectives that are not wholly incommensurate with one another. In other words, there are parts of each perspective that might fit together to form a broader approach on the population-resource relationship.

6.4 Environmental Implications of Population Trends

What do these current and future demographic trends mean for the environment? The IPAT model, which posits that (Environmental) Impact = Population x Affluence x Technology, serves as a useful tool in unraveling the complex relationship between population and the environment.

Thus, if any one of these factors decreases, the overall environmental impact decreases and vice versa. All things being equal, therefore, more people generally lead to greater environmental degradation at all geographic scales, from the global to the local. Of course, in reality it is not that simple, since "all things being equal" is not a realistic assumption in many instances; there are always multiple variables and contingencies involved. Indeed, despite a wealth of information and knowledge available today, there remain few easy answers.

Take for example fairly recent trends in global infant mortality: global infant mortality rates have fallen by over 50% since 1990. From a humanitarian perspective, this is an amazing improvement over a 25- to 30-year period! Demographically, it helps explain increasing life expectancies in developing regions and, by extension, population growth in many developing countries. With infant mortality rates declining around the world, more babies are surviving past the age of one and into adulthood, which, in turn, leads to greater population growth. At the same time, standards of living (affluence according to the IPAT model) are generally increasing in the Global South (although marked by significant geographic unevenness). Nearly everyone in the world would consider these demographic trends good news, a sign of continued progress in the context of global humanity. And yet, there is the rub, as Hamlet famously said. These progressive trends and overall improvements in people's lives, especially in the Global South, would not be seen as "good news" in the context of the IPAT model, since both population and affluence are increasing. Thus, without a significant offset to those changes in population and affluence by technological advances, the earth would suffer increased environmental impact.

So, while the numbers are important, they only tell part of the story. It depends, first, on who is doing what. For example, who has regular access to electricity and how is that energy produced (part of the "A" and the "T" in the IPAT model)? Who has access to clean water, how reliable and consistent is it, and how do they obtain it? Next, it depends upon where; geography is crucial to questions of environmental impact. Despite the seemingly obvious impact of global population growth expected to crest at 11 billion by the end of the twenty-first century, most climate policy recommendations do not directly address population policy as a potential means to mitigate the effects of climate change. Notably, the 2018 report from the United Nations' Intergovernmental Panel on Climate Change (IPCC) stops short of proposing any direct action regarding population growth, although it does point out the ways in which population and climate change affect one another. Although it is true that slowing population growth in high fertility (generally lower income, developing countries) will not make much difference in the short term, the cumulative effects over the course of the twenty-first century would be notable. This is particularly the case given the projected increases in development (and thus GHG emissions) in many developing countries in the coming decades. The argument goes, therefore, that although short-term impacts of significant population growth decline in high-fertility societies may be negligible, the long-term cumulative effects are worthy of consideration in climate change policy circles. Recently, longtime population researchers like John Bongaarts (Bongaarts and O'Neill 2018) have made this argument inurging the IPCC to consider population policy more comprehensively as an additional factor that could aid in both climate change mitigation and adaptation efforts. Bongaarts and O'Neill (2018) further explain that human rights-based population policies hold the potential for multiple improvements to people's lives in developing countries, including improving gender equity and helping to achieve the UN's sustainable development goals.

> This type of population policy generally comes in the form of family planning programs that provide access to affordable modern contraceptives in high-fertility areas of the world.

Although increased access to modern contraceptives does not guarantee women will utilize them, the vast majority of demographic research indicates that women do generally take advantage of increased opportunities to plan the number (and spacing) of births. Additional policy proposals might include improving access to education for women and girls, since increased educational attainment has been found to result in a decrease in fertility rates in nearly every region of the world.

As noted previously in this chapter, any discussion of climate change and population must account for the massive disparity in per capita greenhouse gas emissions between developed and developing countries. For instance, China emits more CO2 each year than any other country (about 10 billion metric tons annually), and the United States emits just over 5 billion metric tons each year. However, the average resident of China is responsible for approximately 7.5 tons of CO2/year, while the average German emits 8.9 tons/year, and per capita emissions in the United States exceed 16 tons/year. By way of comparison, the global average is much lower at 4.9 tons of CO2 per person/year. Thus, for many researchers and policymakers, looking to implement population growth policies in developing countries as a means of addressing climate change amounts to blaming poor countries for the climate crisis. Of course, as we see in the earlier examples, the reality is that it is us in the Global North who deserve the bulk of the blame. The challenge, therefore, is in how to promote such a population policy focused on developing countries (who, in many cases, are not contributing much to the problem) and at the same time convince Global North countries' governments and citizens of the urgency of the climate crisis and our choice either to do nothing and exacerbate it or to make sweeping changes in order to avoid some of the worst-case scenarios.

At this point, you may be thinking to yourself, "ok, now what are we to do?" As noted previously, there are no easy answers. Location, distance, resource bases, and spatial relations are only a few geographic concepts that we draw on in future chapters that help us unpack the complex relationships between humans and the environment. Moreover, in several of the subsequent chapters, we discuss geographic and environmental concepts, tools, and illustrations that will enable you to critically engage with contemporary debates integral to human-environment interactions and decide for yourself what should be done in the future.

6.5 Chapter Summary

Although extensive global population growth stands out as a hallmark of the twentieth century, most of human history saw very slight increases or even decreases in the total number of inhabitants of the planet. Indeed, as we saw in this chapter, some historical moments such as the Medieval Black Death in Europe and the Columbian exchange in the Western Hemisphere were marked by their negative demographic impact. Today, however, the earth's population continues to grow, albeit more slowly than during the mid-twentieth century. Moreover, this increase in global population exhibits a geographically uneven distribution, with much of the future growth expected to occur in developing regions of the Global South. The Demographic Transition (DT) model is an imperfect, yet helpful, framework to assess the current and future state of population growth across multiple countries. The decline in mortality that accompanies advances in science and medicine (Phase 2 of the DT) accounts for a great deal of population growth, yet tends to be followed by a decline in birth rates (Phase 3). This birth rate decline is associated with broader structural changes in economic systems (e.g., a shift away from agrarian lifestyles) and/or increased use of modern-day contraceptives and leads to much lower rates of population growth. Although there are exceptions, most developing countries demonstrate a proclivity to follow along similar demographic paths as developed countries, and thus the rate of global population growth will continue to decline.

Despite this slowing of population growth rate, all projections point to the world's population increasing throughout this century, the environmental implications of which we cannot ignore. Neo-Malthusians, Distributionists, and Economic Optimists offer three contrasting perspectives on the population-resource relationship, each placing greater emphasis on one of three factors: population, consumption, and technology, respectively. Finally, the IPAT model provides a basis for evaluating the impact of population on the environment by including the effects of affluence and technology (the latter possibly, although not necessarily, offsetting negative environmental impacts). Given that the number and geographic distribution of people on the earth's surface affect nearly every environmental issue of our time, understanding the historical and contemporary geographies of population growth remains essential. We expect that this historical and contemporary demographic context, along with multiple approaches to evaluating the environmental implications, will serve as a foundation for engaging in the subsequent chapters of this book.

References

Bongaarts, J. (1978). A framework for analyzing the proximate determinants of fertility. *Population and Development Review*, 105–132.

Bongaarts, J., & O'Neill, B. C. (2018). Global warming policy: Is population left out in the cold? *Science, 361*(6403), 650–652. https://doi.org/10.1126/science.aat8680.

"The Development of Agriculture." The Genographic Project. National Geographic. https://genographic.nationalgeographic.com/development-of-agriculture/.

Denevan, W. H. (1992). *The native population of the Americas in 1492* (2nd ed.). Madison: University of Wisconsin Press.

Ehrlich, P. (1968). *The population bomb*. New York: Ballantine Books.

Hardin, G. (1968). The tragedy of the commons. *Science, 162*(3859), 1243–1248.

Koch, A., Brierley, C., Maslin, M. M., & Lewis, S. L. (2019). Earth system impacts of the European arrival and great dying in the Americas after 1492. *Quaternary Science Reviews, 207*, 13–36.

Lovell, W. G. (1992). Heavy shadows and black night: Disease and depopulation in colonial Spanish America. *Annals of the Association of American Geographers, 82*(3), 426–443.

Mann, C. (2005). *1491: New revelations of the Americas before Columbus*. New York: Knopf.

Scott, S., & Duncan, C. J. (2004). *Return of the black death: The world's greatest serial killer*. John Wiley & Sons.

United Nations Department of Economic and Social Affairs, Population Division. (2017) *World Population Prospects: The 2017 Revision, Key Findings and Advance Tables*. Working paper no. ESA/P/WP/248.

US and World Population Clock. US Census. https://www.census.gov/popclock/

Welford, M. R. (2018). *Geographies of plague pandemics: The spatial-temporal behavior of plague to the modern day*. Routledge, U.K.

Agriculture

Contents

M. R. Welford, R. A. Yarbrough, *Human-Environment Interactions*,
https://doi.org/10.1007/978-3-030-56032-4_7

Learning Goals

After reading this chapter, you will be able to:

- Analyze current agricultural practices and patterns in their historical contexts.
- Examine contemporary geographies of agricultural production and their connections to food security.
- Analyze the environmental impacts of contemporary agricultural practices.
- Evaluate the impact of technological innovations on agricultural systems, including their environmental implications.

Over the past 50 years, technological breakthroughs and economic globalization have drastically altered food production regimes across the globe. Specifically, mechanized agriculture—the use of synthetic inputs and, more recently, biotechnology—has transformed the way food is produced in many countries. Although technological advances in food production have increased the total amount of food produced on earth, food insecurity remains a problem for over 800 million of the earth's residents in 2018. In addition, these changes to food production create significant ecological footprints and oftentimes environmental degradation. In Asia, for example, the environmental impacts of the Green Revolution's use of chemical fertilizers, pesticides, and herbicides, which began in the 1960s, are still seen today. This chapter covers a range of issues related to agricultural production including population dynamics; environmental factors shaping agricultural practices; social, economic, and political factors contributing to food (in)security; and globalized food regimes. In addition, it addresses myriad agricultural practices and their environmental impacts including subsistence agriculture, organic farming, the use of synthetic inputs, biotechnology, and fossil fuel-based agricultural systems.

Industrial Agricultural Plantations—Case Study of Oil Palms in Indonesia

Recall that on a global scale, deforestation contributes more greenhouse gas emissions than transportation (i.e., all cars, trucks, and planes combined). Southeast Asia contributes to those emissions through large-scale deforestation for agricultural production. Specifically, 61% of oil palm production originates in Indonesia and 35% comes from Malaysia, with palm oil comprising 30% of world's vegetable oil production. Between 2000 and 2010, oil palm plantations increased 278% in Kalimantan, Indonesia, expanding to 31,640 km^2. Today, nearly all palm oil plantations (95%) in Indonesia are on Kalimantan or the island of Sumatra. Indonesia plans to double oil palm plantation holdings in Kalimantan and Papua over the next 10–15 years, yet deforestation and forest degradation associated with oil palm plantation expansion are projected to contribute 18–22% of Indonesia's CO_2 emissions by 2020 (Carlson et al. 2012). These tropical wetland forests are among the most efficient carbon sequestering forests in the world, measured as carbon dioxide storage per hectare. Yet, between 1990 and 2005, Sumatra and Kalimantan lost 40% of their remaining lowland forest, and this has led to the loss of 150,000 orangutans, pushing them toward extinction. Deforestation of tropical lowland forest in Thailand for oil palm and rubber plantations caused bird species richness to decline 60%, with insectivores and frugivores experiencing the greatest losses, and as deforestation proceeds, high conservation status bird species with restricted ranges are replaced with low conservation status bird species with extensive ranges (Aratrakorn et al. 2006). In China, rubber plantations act as water pumps, evaporating more water than adjacent tropical forest, storing less soil moisture, and causing water shortages in the dry season (Tan et al. 2011).

What is driving this deforestation and plantation expansion is the value of palm oil which generates $18.6 billion in foreign exchange revenues and represents 12% of Indonesian's exports. Although only contributing to 14% of GDP, the palm oil industry is a significant contributor to Indonesia's rural economy, with 41% of palm oil production generated by small, poor land holders. In 2006, 1.7–2 million people worked in the oil palm plantations.

However, deforestation, forest degradation, and the loss of orangutans have inspired local, regional, and global organizations to try to halt deforestation in Kalimantan. In contrast to shade-coffee production where much of the forest is maintained to provide shade for the sun-intolerant coffee, oil palm plantations cut all the forest down, leaving nothing! Furthermore, soil erosion, soil contamination from subsequent pesticide use, and water pollution downstream of oil palm plantations are significant. In 2015, McDonald's pledged to eliminate deforestation from its entire supply chain, while Dunkin' Brands and Yum! Brands (think KFC, Taco Bell, and Pizza Hut) also committed to stop buying palm oil products (Negin 2015). In early 2020, the Kellogg company committed to stop using unsustainable palm oil due to an online petition initiated by two sisters, Asha and Jia Kilpatrick, aged 12 and 10, respectively, from Bedfordshire, that went viral in 2018. Kellogg will monitor palm oil suppliers to ensure the use of sustainable palm oil and also spend money restoring rainforest to conserve orangutans. Asha and Jia are an excellent example of what individuals using social media can do to effect change. In this case, the Kellogg company, fearful of a loss of revenue and unable to smear the two sisters through a negative media campaign, capitulated to their demands. At a more local scale with the help of Charles Roring, a local ecotourism operator, the villages of Klatomok and Malagufuk in Western Papua reasserted their ancestral right to land by occupying the Klasow Valley, thereby thwarting the attempts of the Indonesian government to auction their land for oil palm plantation development. Today, the communities protect 2779 hectares of lowland rainforest and welcome ecotourists to see their beautiful land and its birds-of-paradise and cassowaries.

Food Security In 2016, for the first time in a decade, the percent of the world's population who struggle with obtaining sufficient dietary energy consumption, that is, getting enough to eat on a regular basis, began to rise. According to the United Nations Food and Agricultural Organization (FAO), global food security has been on the rise in recent years, when measured as the percentage of the world's population struggling with food security (or the prevalence of undernourishment). Indeed, after more than a decade of steady declines in global food insecurity, the global prevalence of undernourishment increased in 2016 and 2017 (see table). Ongoing civil conflicts, climate extremes, and increased climate variability contributed, in large part, to this increase, and impacts from climate change were one of the leading causes of major food crises. Moreover, the negative impacts of climate change are expected to proliferate in the coming years. Not unlike many global phenomena, food insecurity is geographically uneven, with some regions suffering more than others. For instance, most of Africa, South America, and Oceania experienced recent declines in food security (accounting for the global uptick in prevalence of undernourishment since 2015), while much of Asia, Europe, and North America remained stable.

Sometimes referred to as global hunger, this is more accurately described as food security—the ability to meet one's basic food needs on a daily basis.

As the World Food Program explains, "People are considered food secure when they have availability and adequate access at all times to sufficient, safe, nutritious food to maintain a healthy and active life" (▸ https://www.wfp.org/node/359289). Food security is about much more than the supply or availability of food, as it encompasses issues related to access, poverty, infrastructure, transportation, and climate change, to name a few. Perhaps not surprisingly, food security issues are more prevalent in regions with agricultural systems that are more susceptible to drastic changes in rainfall (e.g., floods or droughts) and/or temperatures and with populations that are heavily dependent on agricultural livelihoods. Furthermore, climate extremes and variabilities often exacerbate human-driven conflict. Of course, this affects many countries in the developing world.

In absolute terms, food has become more abundant, but there are major disparities between food supply, food quality, and food demand. Africa is a major exception. Large swaths of agricultural land across Africa are increasingly used for cash-crop production for export with the result that overall food production has declined. Although in southern Morocco, despite government efforts, farmers are retaining anti-risk tactics and are maintaining traditional non-intensive agricultural systems such as leaving land fallow, grazing sheep on arable land following crop production, and using more sustainable primitive horse/mule/ox/cow plowing. Even within the USA, there are food deserts in lower-income urban and rural neighborhoods.

Prevalence of Undernourishment (PoU):
2005–14.5
2010–11.8
2015–10.6
2017–10.9
Source—UN Food and Agriculture Organization of the United Nations (FAO 2018)

7.1 Global Food Systems

Per capita Food Production
Developed countries
1960s – 2300 calories/day/person
1990s – 2700 calories/day/person
Developing countries
1960s – 2400 calories/day/person
1990s – 2500 calories/day/person

7.1.1 Food Waste

Recent studies estimate that the USA wastes between 125 and 160 billion pounds of food each year, and 43% of this total is wasted in the household (Gunders 2017; Harvard Food Law and Policy Clinic 2016). Financially, this is the equivalent of \$1800 of wasted food per household. At the same time, approximately 13% of households (i.e., 42 million people) in the USA struggle with food security at some point each year. This wasted food, therefore, could easily alleviate the food insecurity of tens of millions of Americans each year (USDA). In comparison, the average American wastes approximately 10 times as much the food per year than the average resident of sub-Saharan Africa or Southeast Asia. This massive food waste problem in the USA also results in squandering resources, such that one-fifth of cropland, fertilizers, and agricultural water are lost when the food grown via these inputs is not consumed. Hence, maldistribution and waste lay at the heart of the global food problem, but so too does the changing nature of the global diet. Increasingly the world is eating a US diet, full of salts, fats, and worst of all sugars, especially those in the form of high-fructose corn syrup (HFCSs). Fortunately, the U.S. Department of Agriculture and the U.S. Environmental Protection Agency (EPA) have committed to cutting food waste in half by 2030. Moreover, in 2015, the 193 member states of the United Nations (UN) made the same commitment by signing onto the UN's Sustainable Development Goals.

In 2016, the French government banned large grocery stores from throwing away unsold food that could be donated to NGOs and charities that feed the homeless and those who suffer food insecurities. According to Saltzman and coauthors (2019),, the law is working; each morning, some 2700 grocery stores donate food to charities and NGOs, rescuing 46,000 tons of food each year from which some 226 million free meals are prepared! The 2016 law also inspired the creation of the app Too Good To Go that identifies where cheap food is available from smaller groceries and restaurants. However, although heavy fines are part of the 2016 law, no one has yet been fined.

Recently, we have seen progress in addressing this food waste problem in the form of new companies that are incorporating novel methods while also making a profit. One example is a startup based in Silicon Valley in California that focuses on food spoilage by engineering a means to ensure produce has a longer shelf life ("The Amphiphillic Liquid That Keeps Your Avocados Fresh"). Essentially, scientists have created a product made from plant-derived materials (e.g., lipids and glycerolipids) that is applied to produce after it is harvested and prolongs the life of fruits and vegetables. This effectively keeps more moisture in and more oxygen out and thus prolongs the life of produce by extending spoilage time. Although relatively young, this company has already partnered with major grocery store chains and farmers and its avocados are being sold in stores in the USA. Another startup in the San Francisco Bay Area tackles the food waste problem from a different angle, by upcycling used grain leftover after beer production to produce baked goods ("The Secret Edible Powerhourse Hiding in Beer Waste").

> Upcycling (sometimes called creative reuse) refers to the use of waste, by-products, or unwanted products for new, valuable purposes.

These "spent grains," as they are called, amount to quite a bit after the brewing process is finished; for every gallon of beer produced, you end up with approximately three to four pounds of spent grains. Furthermore, spent grains are high in protein and dietary fiber, making them appealing for use in baked goods, such that several companies around the country have utilized these in their bakery businesses, using them for cookies and energy bars, for instance. In a different geographic context, Marmite (or Vegemite in Australia) salty pastes are another widespread example of upcycling. Both are made from yeast extract, a by-product of beer production. Marmite was invented in 1902 and has been used as a paste on sandwiches and as a flavoring in dishes ever since. It started to gain significant popularity during WWI when Marmite was issued to British and Commonwealth soldiers in their rations. Today, Marmite sales in the UK amount to 11.6 million jars that generate $36 million in sales. These are but a few examples of how food waste can be curbed on a small scale, while also contributing to a business model. Although some local, state, and/or federal policies may also help curb food waste in the USA, it will likely take a combination of private and public initiatives to substantially address the problem.

7.1.2 Green Revolution

> The Green Revolution refers to a package of technological advancements in agricultural production that was developed starting in the 1950s and 1960s with the goal of increasing food security in developing countries. With the financial support of the Ford Foundation and Rockefeller Foundation, the research and innovation was undertaken over several years by a team of scientists, with the most prominent, Norman Borlaug, receiving the Nobel Peace Prize in 1970. This package of new technologies included higher yielding varieties (HYVs) of staple crops such as rice, maize, and wheat that required significant amounts of synthetic inputs, especially fertilizers and pesticides, in the production process. In addition, irrigation systems and mechanization tended to be a part of this new approach to producing food, such that all of these inputs are associated with the Green Revolution.

These Green Revolution techniques began to be exported to developing areas in the 1960s and over the next two and a half decades resulted in large increases in the supply of staple crops produced in many developing countries including Mexico (but see Thresholds chapter for discussion of the Crisis in El Campo) and several Asian nations. Because supply was a major contributor to food security challenges at the time in many developing countries, the Green Revolution was successful in achieving greater levels of food security across the developing world through the 1980s. For instance, wheat production increased 145% in China from the mid-1960s to the mid-1980s, while Thailand nearly doubled its rice yields over the same time period.

This success story did not come without its adverse effects, however. Such an upheaval in agricultural production systems in many developing countries caused disruptions in the agricultural labor market. Those farmers better positioned to access the loans and other means of assistance needed to invest in necessary capital inputs (e.g., mechanization, HYV seeds, and chemical fertilizers and pesticides) benefited from this shift in the means of agricultural production, while many laborers were put out of work by the large-scale mechanization. Then there are the longer-term environmental impacts that continue even now. These include common cases of water pollution and subsequent human health hazards from the run-off of pesticides and fertilizers into local waterways. For example, DDT is a pesticide that has been deemed a probable carcinogen and was banned in the USA in 1972. Soil salinization is yet another ecological problem that has occurred in countries that implemented Green Revolution technologies.

7.1.3 Soil Salinization

> Soil salinization is where salts are concentrated in the upper soil horizons in areas with very low effective precipitation and shallow water tables. This is a natural phenomenon where evaporation at the soil-atmosphere interface exceeds precipitation and water (and its associated salts) within the deeper layers of the soil percolate toward the surface.

This is common in places such as Death Valley. However, these salts are toxic to plants. In semi-arid and arid regions where evaporation exceeds precipitation and Green Revolution technologies have expanded the use of irrigation, soil salinization is common, but rather than being immediately apparent after the first use of irrigation, soil salinization is a cumulative process that takes years of irrigation to impact crop yields. But once soil exhibits high salinity, there is no mitigation—the soil is useless! Irrigation-induced soil salinization is widespread and growing; according to Vengosh (2003), in excess of 60 million hectares or 24% of all irrigated land is at high risk of soil salinization.

Rising Global food production Global food production has increased because of several factors. Cropland area has increased, and the hectares of drained and irrigated fields increased. In addition, increasing capital inputs in the form of mechanized farm implements, fertilizer, herbicide, and pesticide; the development and use of higher-yielding crop varieties; and a shift to more productive crops has occurred. For instance, sugar cane (10–15% sugar content) is replaced with sugar beet (13–18% sugar content) as the world's principal sugar crop. However, the rate of increase in food production is declining from 3.7% annual growth in the 1960s, to 2.5% in the 1970s, to just 2.1% in the 1980s. The emergence of pests resistant to pesticides, contamination of water bodies by fertilizer and pesticide residues, transfer of fertile agricultural land to urbanization, land degradation, aging farmers who are not being replaced, and ethical issues about genetic engineering have all contributed to a slowing in food production. Food aid may also be less than desirable.

Free, low-cost food aid drives down food prices, causing local farmers to lose out. Food aid can become an essential part of national food supply, denying recipient countries self-sufficiency in food production. Yet food aid frequently does not reach those in greatest need living in remote rural areas, a classic case of geographic mismatch. Furthermore, food aid is frequently offered in exchange for political support and is often surplus food that accumulates in developed countries; it either is not appropriate or changes local tastes to more Western diets, leading to unhealthy consequences.

Meat Production and Consumption As globalization continues to expand and intensify and millions of people in the developing world move out of poverty, global consumption of meat continues to grow. From the early 1970s until 2010, global meat production tripled, while the world's population grew by approximately 80%; that means meat production grew at 3.5 times the rate of the global population over those four decades (Xue et. al. 2019). Simply put, on average, we are eating more meat worldwide. But does this matter, and if so, why? The short answer is "yes, it definitely matters." Indeed, most scientists note that the single most significant change someone (especially in the USA, where per capita meat consumption is three times the global average) can make to positively impact the environment is to eat less meat. To place US consumption in context, only one country has higher per capita beef consumption than the USA, and that is Argentina. Given that Argentina has a population of 45 million, while the USA has over 330 million residents, the impact of US beef consumption cannot be ignored. Indeed, agriculture and deforestation account for approximately one-fourth of all global greenhouse gas emissions, and in some regions, deforestation is driven primarily by cattle ranching (e.g., in the Brazilian Amazon). Moreover, the amount of grain consumed by livestock animals as feed would be enough to feed nearly 800 million people (Poore and Nemecek, 2018; Shepon et. al. 2016).

Meat production remains inefficient on several fronts. First, it is inefficient to produce, given that, on average in the USA, 14 calories of inputs is required to yield 1 edible calorie of meat (this is referred to as the feed-to-food ratio). This 14:1 ratio is for multiple categories of meat—poultry, pork, beef, eggs, and dairy; beef is the least efficient with a feed-to-food ratio 33:1, while dairy is the most efficient at 6:1. Second, it is inefficient in terms of environmental impact and nutritional value, as the global production of meat, dairy, eggs, and fish contribute over half of the greenhouse gas emissions in global food production, yet provide only 18% of our calories and 37% of protein consumed by earth's residents (Poore and Nemecek, 2018; Shepon et. al. 2016).

Globally, consumer habits hold great potential to mitigate climate change impacts of agriculture. Specifically, research suggests that shifting to a diet that excludes animal products would cut current greenhouse gas emissions from agriculture into half. The impact in the USA would be even more transformative, given the above-average meat consumption of Americans; a hypothetical shift away from all meat consumption would decrease these emissions by two-thirds. Even less drastic changes like substituting poultry for beef in the USA would produce enough calories to feed more than 100 million additional people (or substantially reduce the carbon footprint of meat production) (Poore and Nemecek, 2018). Of course, these are hypothetical scenarios and such large-scale dietary shifts away from meat are not likely to occur on such a massive scale (at least not anytime soon). Nevertheless, these research findings on the impact of current meat consumption and production practices highlight the vast potential for dietary changes to result in significant improvements to the impending climate crisis, while also pointing out the power of global food consumers.

Deforestation in Brazilian Amazon

Since the year 2000, cattle ranching and soy production have driven deforestation in the Brazilian Amazon, with clearing for beef production accounting for 70% of all deforestation. At 27,500 sq. km^2, deforestation peaked in 2004 and then declined fairly steadily for nearly a decade because of a combination of factors. First, soy production retracted, while beef yields increased, and cattle herd size decreased. In addition, authorities cracked down on illegal deforestation, more lands were protected, and highway paving slowed, which hindered access to agricultural frontier areas. As government protection of these areas continued, and following several national programs enacted in 2009, deforestation again declined, as the risks associated with illegal cutting became even more severe (Nepstad et. al. 2014). Although 2013 saw a slight increase compared to the previous decade, as a whole, deforestation decreased by 70% in the Brazilian Amazon in 2004–2016. This is a monumental accomplishment, for which the government, some industry leaders in Brazilian beef production, and international environmental groups like Greenpeace deserve a great deal of credit.

Recently, however, deforestation has been on the uptick in Brazil's Amazon, and massive fires (in quantity and size) brought renewed international attention to the issue in 2019. For example, data from August 2019 show a 222% increase in deforestation that month compared to August of the previous year. Moreover, deforestation data on the last few months of 2019 were not yet available at the time of this writing, but preliminary data suggest that deforestation rates continued to increase compared to 2018, while foreshadowing an even grimmer scenario for 2020. Much of this recent increase in deforestation is attributable to the policies and rhetoric of Brazil's President Jair Bolsonaro, who assumed the office in January 2019. Bolsonaro has long denied that climate change is being driven by human actions and made it no secret during his campaign that his administration would dismantle environmental regulations and open up protected areas of Amazonia to mining and agriculture (Ferrante and Fearnside, 2019).

Not surprisingly then, upon taking office, the new Brazilian president made sweeping changes to the country's Ministry of the Environment, including abolishing the office overseeing climate change policy and shifting oversight of deforestation to the Agricultural Ministry. The Minister of Agriculture is a so-called *ruralista*—a large landholder who promotes deforestation in the name of agribusiness. Moreover, Ricardo Salles, Bolsonaro's newly appointed environmental minister claims that climate change is not human-driven and has dismissed global warming as harmless. Thus, although Bolsonaro did not dissolve the Ministry of Environment as he promised he would during his campaign, these changes made deforestation laws ineffectual, and paved the way for the massive increase in deforestation in 2019, including the fires that gained global attention (Escobar 2019).

Lest you feel defeated, depressed, and ineffectual, recall that a large majority of this deforestation for agriculture is for the purposes of meat production (primarily cattle grazing and feed production). In addition, US residents are a huge driver of the global demand for beef. Individual actions do matter in such cases, as the aggregate of even a slight decrease in beef consumption in a country the size of the United States can make a considerable difference in Amazonian deforestation.

7.1.4 Scientists and Activists

Like many environmental issues, the key is promoting information and widespread education of how and why something like deforestation is so very problematic. Moreover, any formal or informal education campaign must demonstrate to people exactly how and why this (should) matters (or should matter) to them, how they may be part of the problem (in this case, by being a consumer of beef), and, lastly, what they can do to help improve things.

The last chapter of the text, "Practical Solutions," provides a more in-depth discussion of how to address our planet's myriad environmental problems at a variety of geographic scales, from individual actions to global movements.

Deforestation

Following are a small sample of approaches that hold the potential to help alleviate deforestation:

- Set targets on emissions reductions and set incentives (oftentimes market-based in the form of tax credits or agricultural subsidies in many global North countries). Indeed, in the USA, and increasingly worldwide, the data to measure success in this area exist. We simply need governments to use these data and enact policies that will help reach such targets.
- Data could be analyzed to offer myriad options to producers to meet the new targets (i.e., not just a one-size-fits-all plan that is foisted upon farmers)
- Environmental impacts (and how these have been mitigated) could then be communicated to consumers through a variety of means, including labeling and/or public awareness campaigns that highlight the complete, true cost of food when environmental impact is factored in.

7.2 Biotechnology in Agriculture

Broadly speaking, biotechnology refers to any technique that utilizes organisms to make or modify a product or to improve plants and animals. Applied to modern agriculture, biotechnology generally entails some type of genetic modification, with the result being genetically modified organisms (GMOs).

The technological knowledge and ability to splice a desired gene from one organism and insert it into another organism, thus creating transgenic organisms, has only existed since the 1970s (see Fig. 7.1).

Genetically modified (GM) foods, that is, those that contain any genetically modified organisms, have only existed for a few decades. Indeed, the FLAVR SAVR tomato, which was genetically altered to delay spoilage, was the first genetically modified food approved by the U.S. Food and Drug Administration (FDA) in 1992. Nevertheless, in this short time period, GMOs have revolutionized large-scale, commercial agriculture across the globe to the degree that GM foods are ubiquitous in most grocery stores worldwide. For instance, over 90% of corn grown in the United States is genetically modified, primarily as *bt* corn©, which contains the bacterium *bacillus thuringiensis*. This bacterium contains a protein that kills many insects including the corn borer and thus acts as a "built in" pesticide. Other common GM crops in the USA include herbicide-tolerant (HT) soybeans, HT cotton, and *bt* cotton; similar to corn, GM varieties comprise over 90% of the soybeans and cotton grown in the USA today. These herbicide-tolerant crops are immune to chemical herbicides that frequently are sprayed on fields to kill unwanted weeds. The herbicide-tolerant soybeans (which comprise 94% of the soybeans grown in the USA), therefore, remain unharmed, while weeds are eradicated (along with any other plant life that is sprayed).

Given that 94% of soybeans, 91% of cotton, and 89% of corn grown in the USA are HT varieties, scientists continue to study the environmental impacts of these crops. Several studies have found that herbicide-tolerant varieties frequently lead to greater amounts of herbicide sprayed, and oftentimes more indiscriminately, since farmers do not have to worry about harming their crop. Moreover, the main ingredient in these herbicides, glyphosate, has been found to run-off into ditches and streams in close proximity to these agricultural fields, some in concentrations exceeding that considered safe by the U.S. Environmental Protection Agency (EPA). Finally, some of the most lucrative HT products are produced and sold by the same corporation that produces the herbicide (e.g., Round-Up Ready® Soybeans and Round-Up® Weed and Grass Killer).

Fig. 7.1 An old hedgerow with mature trees between Thurston, Suffolk, UK. @John Welford 2020

Perhaps not surprisingly, US. corporate interests have been key players in the global production and selling of GMOs ever since the US Supreme Court ruled in 1980 that genetically engineered organisms could be patented. Today, transnational corporations, like Bayer who bought biotech giant Monsanto in 2018, invest billions in research and development, and thus understandably expect a return on that investment. Nevertheless, like many environmental issues in the world today, these corporations, the farmers who purchase their products, and their consumers pay little attention to the additional ecological and human health costs (what economists call externalities) that result from such large-scale application of chemical inputs like herbicides.

Global Patterns in GMOs Measured by crop area devoted to GMO cultivation, the US leads the world, followed by Brazil. Argentina, Canada, and India also produce large amounts of genetically modified crops, and these five countries combined account for the vast majority of GM cropland worldwide (over 90%). Not all countries have welcomed the agricultural biotech revolution, however, with several banning GM cultivation. For instance, more than half of European Union (EU) countries, including France, Italy, and Germany, have banned the cultivation of GMOs (but not their importation). This ban on the cultivation of GM crops exists despite the fact that the EU as a whole does not restrict the cultivation or importation of genetically modified organisms. Indeed, the EU allows each individual member state to decide on the fate of GMOs within each country's borders. Only a handful of countries have issued a complete ban on GMOs (including imports), Russia being the largest and most geopolitically prominent.

Labeling GM Foods Unlike the United States where there is no labeling requirement for GM foods, the European Union has strict laws governing the labeling of GM products, both those produced domestically and those imported.

7.3 Some of the Most Vulnerable—Smallholder Farmers

> Smallholder farmers are those who cultivate less than two hectares (just under five acres), and there are approximately 500 million such farmers in the world today.

These small-scale agriculturalists live all over the world, tend to utilize precipitation-fed irrigation, and remain among the most vulnerable to climate change. This vulnerability stems from a combination of factors including poor-quality land, lack of access to technology, and/or difficulty obtaining financial support (e.g., loans). Thus, smallholder farmers oftentimes are the most susceptible to negative impacts of a changing climate, such as food insecurity, with local environmental and/or socioeconomic conditions possibly exacerbating their situations. The significant risks associated with climate change are similar for many of these smallholder farmers, despite the fact that some are focused on subsistence agriculture (e.g., many in South Asia), while others may engage primarily in production for export (e.g., coffee in Central America) (Harvey et. al. 2018).

Smallholder farmers around the world already are experiencing multiple negative effects of climate change and increased uncertainties and vulnerabilities, varying by geographic and environmental contexts. Indeed, increased temperatures damaging some crops, erosion of fertile land, the emergence of new diseases and pests, drought and increased uncertainty regarding water for irrigation, as well as extreme rainfall events and/or flooding have become all too familiar to small-scale agriculturalists worldwide. The implications for food security are direct and dire. In parts of Central American, for example, climate models predict decreased yields in major staple crops like maize in less than ten years if current trends continue (Harvey et. al. 2018). Meanwhile in Bhutan, located in the Himalayas of South Asia, retreating glaciers and altered seasonal precipitation patterns are threatening the water supply needed to maintain agricultural productivity, and drastic temperature changes at higher elevations engender questions of crop suitability (Ngawang and Kumar 2018).

Assisting smallholder farmers in implementing changes intended to increase resilience in the face of climate change requires that policymakers at geographic scales ranging from the local to the national and even international have access to information about the specific impacts on in particular geographic, socioeconomic, and environmental contexts. Most experts recommend augmenting the breadth and frequency of surveys that ask smallholder farmers to detail the exact challenges they are facing regarding climate change and the precise adjustments they have made (or would like to make, given the availability of additional resources). Without such surveys documenting impacts and adaptation strategies, it will remain nearly impossible for governments, NGOs, or international organizations to effectively aid smallholder farmers in their efforts to maintain or improve their livelihoods in the face of a changing climate.

For example, recent surveys in Guatemala, Honduras, Nicaragua, and Costa Rica reveal that the vast majority of smallholder farmers experienced an extreme weather event in the past decade, with over a third of those indicating it led to significant food insecurity, and more than one-fourth reporting household income loss. Despite this common experience of being impacted by climate change, less than half of these farmers had engaged in any adaptation strategy, which is a fairly common trend amongst smallholder farmers. Most researchers conclude that this gap between the number who are impacted by climate change and the number who respond to those impacts reflects the limited capability of most of these agriculturalists to make significant changes to their farming practices (Harvey et. al. 2018). Recall that these smallholder farmers, by definition, cultivate small plots of land, and tend to have little access to technological, agrochemical, and/or financial support. Thus, it should not come as a surprise that they frequently struggle with implementing adaptation strategies in response to the changing climate.

Given these circumstances of most smallholder farmers, you would probably not be surprised that in those instances where adaptation strategies have been pursued, they most commonly fall under the category of "ecosystem-based adaptation" (EbA). As defined by the Convention on Biological Diversity, ecosystem-based adaptation is "the use of biodiversity and ecosystem services as part of an overall strategy to help people adapt to the adverse effects of climate change." (CBD 2009). An example of EbA is found in Central America, where the most common strategy utilized by both coffee and maize farmers was planting more trees on farms. These trees aid farmers in by providing crops shade from increasing temperatures and helping to combat effects of extreme rainfall events. Unfortunately, as is the case in Central America, smallholder farmers may also engage other types of adaptation, such as augmenting the use of agrochemicals as short-term fixes that are not sustainable (Harvey et. al. 2018).

7.3.1 Climate-Smart Agriculture (CSA)

According to Lipper et al. (2014), climate-smart agriculture (CSA) "identifies synergies and trade-offs among food security, adaptation and mitigation as a basis for informing and reorienting policy in response to climate change" (1068).

> CSA is a multi-scalar approach that looks to both local and global actors to plan and implement changes to agricultural production strategies that take climate change into account and, by extension, are likely to increase the resilience of both food systems and livelihoods.

Following from IPCC recommendations, Lipper et al. argue that CSA pathways would lead to greater food security and resilience in the face of multiple environmental and socioeconomic stressors, including climate change, while non-CSA approaches do the opposite—they decrease both food security and resilience.

CSA pathways need not address every one of these three objectives simultaneously in all geographic contexts. Nor should local and national governments, NGOs, and other actors be concerned with a "one size fits all" model being forced upon them from on high. Indeed, CSA recognizes the geographically specific agricultural implications of climate change and the myriad paths that stakeholders (from the local to the national scale) might take to meet one or more of the three objectives described earlier. CSA promotes sharing of data-driven, evidence-based strategies that might include any of the following: "integrated crop, livestock, aquaculture and agroforestry systems; improved pest, water and nutrient management; landscape approaches; improved grassland and forestry management; practices such as reduced tillage and use of diverse varieties and breeds; integrating trees into agricultural systems; restoring degraded lands; improving the efficiency of water and nitrogen fertilizer use; and manure management" (Lipper et al. 2014; 1069).

Climate-Smart Agriculture

A CSA pathway meets the overarching goal of utilizing sustainable agricultural practices to increase food security, while incorporating local-scale appropriate climate change adaptation and mitigation strategies.

Three distinct objectives outlined to meet this need include:

1. Improving sustainable agricultural productivity to aid in enhancing income and development
2. Incorporating climate change adaptation and mitigation strategies at multiple geographic scales (from farm to subnational and even national scales) that aid in improving resilience
3. Devising plans to decrease GHG emissions from agricultural (compared to current levels/practices)

In addition, other strategies may include the increased use of data and technology to build resiliency, including the widespread in many developing countries and smallholder farms. Other strategies might entail broader changes in agricultural production (e.g., shifting to different crops or from livestock to crops) or in household/livelihood structure (e.g., increasing non-farm income to offset potential agricultural losses, thus decreasing risk and uncertainty).

Overall, CSA holds great promise, yet there are several challenges in overhauling agricultural and food systems to achieve one or more of CSA's three objectives. First, there is a scalar mismatch of sorts when it comes to the most recent research on climate change (including impacts, mitigation, and adaptation) and the scale at which most strategies would need to be implemented. Much of the scientific research on climate change focuses on the global scale, while CSA approaches must be incorporated at local scales, likely with oversight, and organizational (and financial) support from the national scale. This presents a significant challenge to policymakers and farmers and thus is one of the areas of greatest need for future research.

Second, in order for the state to provide organizational and/or financial support to farmers (particularly smallholder farmers), its capacity to do so must be strengthened. This is crucial in developing countries, yet more developed countries have room for improvement here as well. This might entail improving and enhancing information sharing capabilities, the use of information and communications technologies (ICT) among farmers, and access to inputs such as seeds and fertilizers.

Finally, much of what is outlined as part of a CSA pathway requires financial resources to implement across broad geographic and temporal scales. Not everything is about money, but for such a large-scale, multifaceted, and long-term approach to yield benefits at regional and global scales, there must be financial commitment from national governments across the globe. If countries around the world work collectively to address these three concerns, CSA pathways hold great potential to both increase food security and decrease risks and uncertainties associated with climate change. Climate smart agriculture could indeed save lives in the short term, while helping save the planet in the long term.

7.3.2 Fair Trade

> Fair trade is an approach that seeks to incorporate social and environmental responsibility into the commodity chain for certain consumer products (e.g., agricultural products like coffee, tea, and chocolate, as well as some textiles and clothing). Most fair trade organizations provide fair trade certifications to farmers and other producers and work as an intermediary between the farmers and retailers (i.e., the sellers).

Through this certification process, fair trade helps to ensure that workers (e.g., farmers and farmworkers) receive a fair wage and that sustainable agricultural practices are maintained, among other production standards. In general, large fair trade certification groups (e.g., the largest in the world Fair Trade USA®) is to provide certifications to producers, provide them a premium or minimum price (usually paid by retail sellers), and thus to inform consumers that these products are produced with a commitment to social and environmental responsibility. In short, it is a way for consumers to know that their coffee or chocolate was produced according to a set of standards that ensure fair wages, working conditions, and sustainable agricultural practices. Although coffee was one of the first products to receive fair trade certification, the fair trade movement has found its way into the production of many agricultural products and even some clothing, with the involvement of major multinational corporations like Starbucks, Walmart, Target, and Patagonia.

Arable Hedgerows

According to the Royal Society for the Protection of Birds (RSPB), hedges (or hedgerows) may support 80% of the UK's woodland birds, 50% of the UK's mammals, and 30% of the UK's butterflies (RSPB 2020).

Hedgerows provide extremely valuable ecosystem services that, according to Forman and Baudry (1984), include reducing wind speed, evapotranspiration, and soil desiccation that reduce soil erosion and nutrient loss. Hedges also aid pest control in adjacent fields. Morandin and coauthors found that predatory ladybird bugs (beetles) were more common and aphids less common in arable fields with hedgerows. These benefits extended up to 100 meters from the hedges. In semi-arid Kenya, Girma and coauthors (2000) also found that hedges, especially surrounding fields of maize, provide significant pestconsumption benefits especially due to the presence of parasitic wasps. Hedges also provide essential wildlife corridors that often link both small and large woodlands. Hedges also provide human resources: fruits, seeds, and bushmeat for human consumption; firewood; and enhance the landscape aesthetics of arable farmland across Europe. The lack of hedgerows in the American Midwest certainly contributes to the lack of animal and plant diversity in this region and makes the landscape appear quite barren. However, Morandin and coauthors show that even in simplified agricultural landscapes, the cost of planting hedgerows is recouped through agricultural pest control within 16 years or seven if native bee pollination is also calculated.

Allotments

A facet of the European landscape for a 100 -years or more, allotments are small plots of land rented by individuals for the non-commercial gardening or growing of food. Encouraged during WWI and WWII, these plots still dot the European rural, urban, and suburban landscape, and many males and families retreat to them during the weekends to relax and socialize and tend their crops. In the UK, urban allotments remain popular, and waiting lists for their use are long (Figs. 7.2 and 7.3).

Their equivalent in the USA are community gardens. Wood and coauthors (2016) suggest the UK urban populations unconsciously appreciate their benefits, as allotments appear to play a key role in promoting mental wellbeing. Sadly, as much as 35% of allotment sites have disappeared in the last two to three decades as developers have swooped in and purchased prime residential and commercial sites.

Fig. 7.2 An end-of-garden allotment in continuous production since 1973, and in the background, an ancient hedgerow in Elmswell, Suffolk, UK. @ John Welford 2020

This decline is quite disturbing as allotments and community gardens play a vital role in connecting people to the land and the process of food production and while enabling allotments and community gardeners to supplement their food in a sustainable and cheap manner.

Fig. 7.3 A traditional suite of allotments just outside of Bury St. Edmunds. @ John Welford 2020.

Chapter Summary

We live in an increasingly globalized world, and the ways in which most of earth's inhabitants meet (or fail to meet) their everyday food needs is in some way dependent upon other people and places. Regarding supply, the amount of food produced on a global scale is more than enough to achieve 100% food security today. The problem, therefore, is not one of supply, but one of access. Moreover, issues related to access (e.g., poverty, interruptions stemming from human conflicts, or natural disasters) will continue and likely worsen, given global population projections and probable scenarios associated with climate change. Advances in agricultural production throughout history are associated with hugely significant moments in human history. These range from the earliest domestication of plants and animals in ancient Mesopotamia, as well as Nile, Indus, and Yellow River, enabling these sedentary civilizations to thrive, to mid-twentieth-century Green Revolution farming that helped feed tens of millions in developing countries. More recently, agricultural biotechnology and its resulting genetically modified foods have increased yields in many cases. Nevertheless, these recent advancements have come with massive costs, many of which are environmental in nature. The Green Revolution's dependence on synthetic pesticides, herbicides, and fertilizers continues to run off into and contaminate local water supplies in many global south locales, polluting drinking water sources with known carcinogens. At the same time, meat consumption (especially beef) is primarily responsible for large-scale deforestation in South America, with much of this demand originating in Global North countries like the United States. Some argue that technology will save us, pointing to advances in genetically engineered agriculture as the wave of the future. Others point out that the current corporate-driven globalized food regime has no interest in addressing environmental issues, from local deforestation and water pollution, to global climate change, and thus that any appeal to a continuation of "business-as-usual" will only lead to exacerbating an already grim environmental picture. For those with such well-founded concerns, organic farming (both small and large scale), community gardens/allotments, and the global fair trade movement are appealing alternatives to today's globalized food system. The question that remains, however, is how much impact individuals, government, and/or corporate stakeholders will be willing to make in the future (i.e., how many people need to commit to individual changes and how many corporate and government entities will make large-scale changes). We address some of these potential solutions in the book's concluding chapter.

References

"The Amphiphilic Liquid Coating That Keeps Your Avocadoes Fresh.". *Wired.* https://www.wired.com/story/apeel/

Aratrakorn, S., Thunhikorn, S., & Donald, P. F. (2006). Changes in bird communities following conversion of lowland forest to oil palm and rubber plantations in southern Thailand. *Bird conservation international, 16*(1), 71–82.

Carlson, K. M., Curran, L. M., Ratnasari, D., Pittman, A. M., Soares-Filho, B. S., Asner, G. P., Trigg, S. N., Gaveau, D. A., Lawrence, D., & Rodrigues, H. O. (2012). Committed carbon emissions, deforestation, and community land conversion from oil palm plantation expansion in West Kalimantan, Indonesia. *Proceedings of the National Academy of Sciences, 109*(19), 7559–7564.

CBD (Convention on Biological Diversity) (2009) "Connecting biodiversity and climate change mitigation and adaptation: Report of the second ad hoc technical expert group on biodiversity and climate change." *CBD Technical Series No. 41.* Montreal: Convention on Biological Diversity.

Escobar, H. (2019). Brazil's deforestation is exploding—and 2020 will be worse. *Science Magazine. American Academy of Arts and Sciences.* https://doi.org/10.1126/science.aba3238.

FAO, IFAD, UNICEF, WFP and WHO. (2018). *The state of food security and nutrition in the world 2018. Building climate resilience for food security and nutrition.* FAO: Rome.

Ferrante, L., & Fearnside, P. M. (2019). Brazil's new president and 'ruralists' threaten Amazonia's environment, traditional peoples and the global climate. *Environmental Conservation, 46*, 261–263. https://doi.org/10.1017/S0376892919000213.

Forman, R. T., & Baudry, J. (1984). Hedgerows and hedgerow networks in landscape ecology. *Environmental Management, 8*(6), 495–510.

Girma, H., Rao, M. R., & Sithanantham, S. (2000). Insect pests and beneficial arthropods population under different hedgerow intercropping systems in semiarid Kenya. *Agroforestry Systems, 50*(3), 279–292.

Gunders, Dana et al. (2017). Wasted: how america is losing up to 40% of its food from farm to fork to landfill. National Resources Defense Council. https://www.nrdc.org/sites/default/files/wasted-2017-report.pdf

Harvard Food Law and Policy Clinic. *Federal Enhanced Tax Deduction for Food Donation: A Legal Guide* (2016). http://www.chlpi.org/wp-content/uploads/2013/12/Food-Donation-Fed-Tax-Guide-for-Pub-2.pdf

Harvey, C. A., Saborio-Rodríguez, M., Martinez-Rodríguez, M. R., Viguera, B., Chain-Guadarrama, A., Vignola, R., & Alpizar, F. (2018). Climate change impacts and adaptation among smallholder farmers in Central America. *Agriculture & Food Security, 7*(1), 57.

Lipper, L., Thornton, P., Campbell, B. M., Baedeker, T., Braimoh, A., Bwalya, M., Caron, P., Cattaneo, A., Garrity, D., Henry, K., & Hottle, R. (2014). Climate-smart agriculture for food security. *Nature Climate Change, 4*(12), 1068–1072.

Morandin, L. A., Long, R. F., & Kremen, C. (2014). Hedgerows enhance beneficial insects on adjacent tomato fields in an intensive agricultural landscape. *Agriculture, Ecosystems & Environment, 189*, 164–170.

Morandin, L. A., Long, R. F., & Kremen, C. (2016). Pest control and pollination cost–benefit analysis of hedgerow restoration in a simplified agricultural landscape. *Journal of Economic Entomology, 109*(3), 1020–1027.

Negin, E. (2015). McDonald's pledges to eliminate deforestation from its entire supply chain. Huffington Post. 04/21/2015. https://www.huffingtonpost.com/elliott-negin/mcdonalds-palm-oil-pledge_b_7104982.html. Accessed 10/01/2018.

Nepstad, D. C., McGrath, D. G., Stickler, C., Alencar, A., Azevedo, A., Swette, B., Bezerra, T., DiGiano, M., Shimada, J., Seroa da Motta, R., Armijo, E., Castello, L., Brando, P., Hansen, M., McGrath-Horn, M., Carvalho, O., & Hess, L. L. (2014). Slowing Amazon deforestation through public policy and interventions in beef and soy supply chains. *Science, 344*, 1118–1123.

Ngawang, C., & Kumar, L. (2018). Climate change and potential impacts on agriculture in Bhutan: A discussion of pertinent issues. *Agriculture and Food Security, 7*, 79–91.

Poore, J., & Nemecek, T. (2018). Reducing food's environmental impacts through producers and consumers. *Science, 360*(6392), 987–992.

RSPB. The value of hedgerows for wildlife. Accessed 17 Mar 2020. https://www.rspb.org.uk/our-work/conservation/conservation-and-sustainability/advice/conservation-land-management-advice/farm-hedges/the-value-of-hedgerows-for-wildlife/

Saunders, L. E., & Pezeshki, R. (2015). Glyphosate in runoff waters and in the root-zone: A review. *Toxics, 3*(4), 462–480.

Saltzman, M., Livesay, C., Martelli, J., & Gouffran, D. (2019). Is France's groundbreaking food-waste law working? PBS. https://www.pbs.org/newshour/show/is-frances-groundbreaking-food-waste-law-working

The Secret Edible Powerhouse Hiding in Beer Waste." *Outside Online* https://www.outsideonline.com/2070151/secret-edible-powerhouse-hiding-beer-waste

Shepon, A., Eshel, G., Noor, E., & Milo, R. (2016). Energy and protein feed-to-food conversion efficiencies in the US and potential food security gains from dietary changes. *Environmental Research Letters, 11*(10), 105002.

Tan, Z. H., Zhang, Y. P., Song, Q. H., Liu, W. J., Deng, X. B., Tang, J. W., Deng, Y., Zhou, W. J., Yang, L. Y., Yu, G. R., & Sun, X. M. (2011). Rubber plantations act as water pumps in tropical China. *Geophysical Research Letters, 38*(24), 1–3.

United Nations World Food Program. www.Wfp.org

United States Department of Agriculture (USDA) Food Access Research Atlas. https://www.ers.usda.gov/data-products/food-access-research-atlas/go-to-the-atlas/

Vengosh, A. (2003). Salinization and saline environments. *Treatise on geochemistry, 9*, 612.

Wood, C. J., Pretty, J., & Griffin, M. (2016). A case–control study of the health and well-being benefits of allotment gardening. *Journal of Public Health, 38*(3), e336–e344.

Xue, L., Prass, N., Gollnow, S., Davis, J., Scherhaufer, S., Östergren, K., Cheng, S., & Liu, G. (2019). Efficiency and carbon footprint of the German meat supply chain. *Environmental Science & Technology, 53*(9), 5133–5142.

Urbanization

Contents

M. R. Welford, R. A. Yarbrough, *Human-Environment Interactions*,
https://doi.org/10.1007/978-3-030-56032-4_8

Learning Goals

After reading this chapter, you will be able to:

- Define urbanization and explain what recent urban trends behold for the future of the globe.
- Analyze urban growth and discuss the future of cities in the context of global environmental challenges.
- Assess the impact of cities on the world's climate crisis now and into the future.
- Evaluate measures cities are taking to address contemporary environmental challenges.

In 1950, a mere 30% of the world's population was urbanized. By 2020, that number had climbed to over 55%, and projections indicate that by 2050 over two-thirds (approximately 68%) of the world's residents will live in cities since the world currently is experiencing an annual urbanization rate of 1.9%. Looking more closely at the world's 4.1 billion urban residents, over half live in smaller cities of less than one million, and geographically, there remains a great diversity in levels of urbanization across world regions; North America and Latin America/Caribbean are more than 80% urbanized, while about three quarters of Europeans live in cities. Approximately two-thirds of Oceania's residents live in cities, and Asia as a whole is just at 50% urbanized. Africa is the least urbanized region in the world, with 43% of residents living in cities. Cities in the Global South are growing much faster than those in the Global North, and strikingly, nearly 90% of this 2.5 billion increase in the world's urban residents by 2050 will occur in Asia and Africa. One specific example illustrates this future trend: Tokyo has been the world's most populous city for decades (37 million in 2020), but its growth is slowing, and contraction is imminent. Therefore, in the near future, Delhi, India, is expected to become the most populous city in the world, as it will add a projected 8 million residents to surpass the 37 million mark by 2028! Although we may immediately think of megacities like Tokyo (Japan), Shanghai (China), Mumbai (India), and New York (USA) when picturing global urbanization, it is important to note that only 25% of the earth's residents live in cities with more than one million residents.

> Distinguishing between the level and rate of urbanization is helpful: level of urbanization refers to the simple proportion of a population that resides in urban areas (e.g., the 55% global urbanization level in 2020). The rate of urbanization, by contrast, is the degree to which that number is changing (generally, this would be an increase, although theoretically it could decrease in any given geographic context). So, the level is the share of a population living in cities, while the rate is how fast that number is increasing.

Perhaps not surprisingly, Africa has the highest rate of urbanization in the world, driven by significant and ongoing rural-to-urban migration. Indeed, rural-to-urban migration is the most prevalent cause of urbanization in the world today, and will continue well into the future, primarily in the Global South. Nevertheless, two other factors could potentially cause urbanization to increase. First, a simple increase in populations in cities through birth rates, without a commensurate increase in rural populations, would result in a higher level of urbanization in that country or region. Although this is technically possible, it is not at all common today. Second, a country's level of urbanization might increase when an urban area expands geographically to incorporate land and populations that were not previously part of the urban area. This sprawl effect could then lead to greater urbanization, since a larger share of the country's population would be living in cities. Unlike the most common cause of rural-to-urban migration, however, wherein the people move to the city, this sprawl entails the city effectively coming to the people.

The substantial growth predicted in urbanization in the developing world presents immense challenges regarding housing, employment, transportation, and infrastructure (e.g., access to potable water and basic sanitation), particularly for newly arriving urbanites. Somewhat paradoxically, the cities and new urban residents that will struggle the most are those very same ones that are least equipped to meet these emergent needs. Thus, sustainable development in Global South cities is crucial to meeting the global environmental challenges of the twenty-first century, as urban areas in developing countries experience a pace of urbanization rarely seen in human history. If we are to avert environmental disaster, there is no option other than managing urban growth in developing countries in a way that focuses on the needs of new urban residents in the coming decades. Simply put, cities must be part of the solution to our impending global environmental crises.

8.1 Sprawl

Effectively, sprawl (sometimes referred to as urban sprawl) describes an urban area that is spread out, especially as you get farther and farther from the urban core, which tends to be very densely populated. Sprawl is characterized by low population density, a lack of street connectivity, and segregated development, oftentimes resulting from a dearth of urban/transportation planning.

This geographically dispersed ("sprawling") urban form tends to stem from areas being designated as single land use (aka segregated development), so that where someone works is separated from where someone lives, shops, and spends her free time. Adding to this segregated development, sprawling cities generally have very few transportation options other than driving a car to meet one's daily needs because these either do not exist, are not a safe alternative, or do not go where people want to go. The USA is the country most frequently associated with sprawl, primarily in a post-WWII context, as cities' populations boomed along with suburban expansion, facilitated, at least in part, by construction of the interstate highway system that began under President Dwight D. Eisenhower in 1956. Thus, sprawl is the USA is auto-centric, since to the extent to which it was planned, development is designed around cars as the primary mode of transport.

This lack of planning and coordination between alternative forms of transportation (something other than a car) and urban development is a hallmark of the built environment of a sprawling urban area. People simply have few alternatives to meet their day-to-day needs other than getting in the car and driving, oftentimes over very short distances. Thus, the availability and affordability of automobiles in the USA in the mid-twentieth century helped to facilitate the expansion of suburban areas; the predominance of cars have been a major impediment to urban planning efforts. The environmental implications for this geographic expansion of urban areas are as detrimental as they are obvious. Traffic congestion, pollution, increased energy consumption, and greenhouse gas emissions, along with negative health effects like asthma and obesity (adults and children), can all be traced to sprawl in the USA. Fortunately, there are efforts in cities across the USA to promote greater connectivity of streets and improvements to the built environment (e.g., expanded sidewalks and bike lanes), along with enhanced and expanded mass transit options that can help combat these decades-long impacts of the USA's sprawling cities.

> New urbanism entails smart growth in the form of high-density, mixed-use developments and alternative transportation options (like bicycling, walking, or taking public transit); it emerged in the 1990s and has affected sprawl in cities across the United States.

Recent studies indicate that sprawl peaked in the mid-1990s and has slowly decreased since then, when measured by street connectivity. These studies stress that the environmental gains from growth policies (mostly enacted at the local scale) and decreased sprawl might be long-term gains on the order of decades, paralleling the time scale on which negative impacts of sprawl were experienced. Although not all urban areas in the USA reflect this positive trend in urban form and expanding transportation options (e.g., one author's hometown of Atlanta, GA), the country as a whole is experiencing less and less sprawl over time. In addition, researchers expect these trends to continue, and for the positive impacts on greenhouse gas emissions and air pollution, to name just a few, to compound over the coming decades, and possibly the next century.

8.2 Urbanization Versus Nature

Although sprawl certainly has encroached upon nature in much of the world, nature in some ways is fighting back! This fight back is frequently detrimental to human health, but urban dwellings, green and brown spaces, and urban utilities such as water and sewage lines, telegraph poles, and wires can and do offer new habitats to opportunistic animals, plants, and pathogens. Of the eight "horsemen of the apocalypse" ready to expose the world's population to new deadly human-to-human vectored diseases—increases in population size and urban densities, the failure of health programs and the emergence of hospital-acquired infections, contaminated water supplies, and persistent poverty (especially in expanding peri-urban slums) come cap-in-hand with urbanization. Many human pathogens require densely packed areas of human habitation to circulate and kill human hosts through pneumonic/blood/sexual contact, human-to-human pathways or vector-driven transmission, or the consumption of contaminated water, plants, or animals.

Today, dengue fever is more prevalent and persistent in tropical urban settings than in rural areas because its vector, the *Aedes aegypti* mosquito, that reinfected the Americas after 1974, breeds in the multitude of stagnant pools of water (e.g., discarded car tires, exterior plant pots, blocked roof gutters) that dot the urban landscape. In 1986–1987, over a million infections are estimated to have occurred just in Rio de Janeiro. However, government-assisted investment in window screens and public awareness programs to repair damaged screens and reduce stagnant pools of water in urban areas in Dominican Republic has curtailed infections in the DR.

Following the Haiti earthquake of 2010, cholera erupted in Hispaniola, infecting 800,000 and killing over 9000 in Haiti. Initially constrained to the Artibonite River drainage, insufficient water and sanitation infrastructure, coupled with a massive earthquake in 2010, made Haiti particularly vulnerable to a waterborne disease epidemic. Rather surprisingly, cholera was subsequently recorded in the USA and was concentrated in populations self-identifying with Hispaniola. In other words, US cholera outbreaks were limited to urban populations of Hispaniolan ancestry where individuals or groups had volunteered to assist in the post-earthquake environment providing humanitarian aid in Haiti. Sadly, the initial source of this cholera epidemic was an individual or individuals among the UN peacekeepers from Nepal.

Socioeconomic spatial inequalities within and between urban and suburban neighborhoods and rural areas create areas and/or regions that support higher pathogen loads. Tuberculosis (TB), unknown in pre-history, thrives in poor, densely populated neighborhoods where many humans suffer compromised immune systems and comingle at much higher rates than rural peoples. The implementation of colonial taxation, arable land acquisition, and food price control in the Republic of Congo and Congo triggered, in the early 1900s, wholesale adult male migration into Brazzaville and Kinshasa. Where once prostitution was uncommon, it expanded to cater for the presence of large numbers of single male displaced from their rural homes. Subsequently, the first known cases of AIDS were identified between 1959 and 1960 in Kinshasa, although it is highly likely that AIDS existed in the rural populations of the Republic of Congo, Congo, and Cameroon for hundreds of years before 1959/1960, infecting just one or two people and then dying out due to a lack of both comingling and prostitution. These socioeconomic spatial inequalities exhibited in west-central Africa in the early 1900s were mirrored in Malawi more recently where large numbers of miners contracted both TB and AIDS and then transmitted these diseases to their domestic partners. Malawi's entrenched poverty, dependence on mining, and urbanization of mining regions maintain this persistent transmission pathway.

The world's biggest mass murder, *Yeresina pestis*, pneumonic plague killed over 75 million in Europe in 1346–1815. In the UK, the highest death rates attributed to plague during the Medieval Black Death in 1348 were associated with events of highest comingling such as urban markets, festivals, and pilgrimages, while killing millions more in Central Asia during this time. In its modern bubonic form, since erupting from China in 1894, plague has killed at least 25 million across the world. In Madagascar, highland black rats continue to act as a reservoir for plague, and periodically transmit their plague to urban/suburban rats, precipitating significant and lethal outbreaks of plague. During the Indian Plague of 1896–1898, plague deaths were nearly entirely restricted to urban and suburban areas and railway depots that had large grain stores. In particular, people living in poorly sealed domestic housing where black rats could access roofs and attics suffered significantly higher mortalities than the rich and those living in rural areas. To this day, Indians flee urban areas when plague is discovered and relocate frequently for the duration of the epidemic to rural areas. Since 1894, bubonic plague has spread to much of the world except Europe. The USA was first infected in 1900 in San Francisco. Today, plague is spreading across the western US as a wave at about 80 miles/decade. Today, urban rat eradication programs minimize urban disease risk in the USA just as similar programs in Brisbane and Sydney, Australia, in 1900, eliminated plague after eradicating urban rat populations. However, between 1 and 17 cases of plague still occur in the USA each year, but these are mostly isolated incidents and occur in remote rural areas. Another highly lethal although rare pathogen is the Hantavirus. The Hantavirus is spread by rodents through their urine, and tends to occur after mild winters whereby rural rodent populations undergo a boom cycle in populations. In the subsequent fall and early winter, large numbers of rodents invade urban and suburban environments seeking food and bringing the Hantavirus with them. In 1993–2018, typically 25–30 people were infected each year, with 5–8 dying from the disease.

Several of the most feared emergent diseases are swine (H1N1 flu strain) and bird flu (H5N1 flu strain). The periodic eruption of swine flu (e.g., the 2009 global epidemic) is tied to the intense comingling in China of people and pigs, where pigs are kept for consumption in urban/suburban and rural backyards. In contrast to swine flu, bird flu's reservoir is in wild birds; however, many birds, especially ducks and geese, are kept in urban/suburban and rural backyards across China. The fear is that if bird flu crosses into domestic ducks and geese in China, human infection with the highly lethal H5N1 strain would occur quickly afterward. Although the global media's response to the 2009 swine flu epidemic was hysterical and panic-inducing, the 2009 strain was not very lethal but very infective. Whereas it is a widely held scientific belief that bird flu would be very lethal and highly infective.

■ Suburbanization

Suburban lawns are a very visual, very recent environmental catastrophe that shine a very poor light on our relationship with nature. Lawns are biological deserts;

furthermore, the millions of gallons of chemical fertilizers and pesticides and herbicides that are sprayed on them, in particular in North America, each year are very significant point-sources of pollution in the suburban and downstream environments.

The emergence of lawns occurred in the 1950s as suburban sprawl and personal car ownership exploded. In the USA, the purchase of and subsequent shutdown of public transportation systems by Chrysler, Ford, and GM led to more and more car ownership.

Moreover, in the USA, although average house sizes did not increase for four decades until the 2000s, lot (parcel) size doubled. In contrast, in many European suburbs, both front and back gardens are positively festooned with trees, shrubs, rockeries, and home-grown food. In urban and suburban areas across Europe, many front lawns have been converted to gravel/tarmac/concrete parking, reducing pesticide and fertilizer applications but increasing urban water runoff and flash-flooding. The front lawn in ◻ Figs. 8.1 and 8.2, the home of one coauthors' parents in Suffolk, England, is not an uncommon sight.

Again, in contrast, Americans overspray and overwater their lawns, and their lawns require significant human labor and gasoline to maintain. For instance, 10% of Delaware is now lawn (this includes residential and commercial lawns and golf courses)! As an irrigated crop, grass acreage is bigger than the next eight combined irrigated crop acreages! Lawns in the USA also consume four times more water than any other single irrigated crop. According to *People, Places and Plants*, a typical US family living on a third of an acre consumes five gallons of gas mowing and trimming, applies seven gallons of oil-based fertilizer, consumes through electrical generation five gallons for watering, and one gallon for other lawn maintenance for a grand total of 18 gallons each year. With 127 million households in the US in 2019 households in the USA, that equates to 2.286 billion gallons of gasoline or 44 billion pounds or 21.7 million tons of CO2 per year.

Sadly, in the USA, a well-kept lawn is a sign of status. It is a simple equation—lawns cost money. A well-kept lawn translates into higher property and resale values; it also tells people you are a good neighbor. As a result, lawn care is a multibillion-dollar business in the USA; in contrast, in Europe, but particularly in the UK, garden centers/nurseries are a multibillion business. The typical garden

◻ **Fig. 8.1** Over a 100-year period, homes in the UK went from terrace with no gardens (built in the 1830s Waterloo, London), to terrace blocks with a two small gardens (built in the 1930s Windmill Gardens, Enfield), to individual houses with substantial front and back gardens (built in the 1972 Elmswell, Suffolk) @Welford 2010

Fig. 8.2 The Welfords' front lawn is a dry garden. Although Mr. Welford will water the scrubs occasionally, he rarely applies either fertilizer or pesticides. (Mistral, Elmswell @ Welford 2010)

8

centers/nurseries cater to those, the majority, who plant trees, scrubs, flowers, and crops, and put in rockeries (i.e., a traditional rockery is an arrangement of rocks and alpine plants) and ponds in their gardens. For instance, The Chelsea Flower Show attracts hundreds of thousands of visitors and many film crews to catch the latest and greatest in designed gardens.

In the USA, grass wins because nothing should impede the view of the house; in contrast, in Europe and the UK, both the garden and the house count toward curb appeal. A simple maintained lawn is not sufficient. It is considered a sign of laziness and a lack of house pride. People genuinely feel that a well-crafted and maintained garden that is pleasing to the eye is a sign that the interior of the house will also be well-maintained. It is quite possible that extending the house-proud feel to the garden among the working and middle classes has its origins in the huge formal and informal gardens that the likes of Capability Brown created for "country" houses, those current and former residences of the landed gentry. So, what was begun by the likes of Capability Brown has been organically embraced by both the working and middle class. In other words, you might live in a terrace house, but you still try to improve its curb appeal by its garden. In the end, the British are garden snobs, while US suburbanites are lawn snobs! Both are engaged in establishing curb appeal but from two very different perspectives with two very similar outcomes.

For instance, suburban gardeners in the UK and Europe still use neonicotinoid chemicals, although European farmers were banned from using them in 2013. In fact, 50% of all pollen and nectar samples collected from bees in the UK in 2018 were contaminated with neonicotinoids. So, although banning all neonicotinoids from farmland in 2013 has had a very positive impact on rural bee populations, the same cannot be said for suburban bees. Neonicotinoids were identified among bug sprays and contaminated ornamental plants sold in garden centers and nurseries (see Nicholls et al. 2018). Sadly, in 2019, neonicotinoids are still widely used in the USA.

According to O'Neill and the National Audubon Society, sprawl in states among Texas, California, New Mexico, North Carolina, Georgia, Alabama, and Florida, coupled with climate change, will contribute to the following species becoming vulnerable to local extinctions: Allen's hummingbird, black oystercatcher, California thrasher, red-cockaded woodpecker, roseate spoonbill, fish crow, swallow-tailed kite, and white-crowned pigeon. Interestingly, the North American wood stork, whose traditional NA range was restricted to Florida and Alabama, has over the last 30 years expanded its breeding range in Georgia and South Carolina. Any discussion of the built environment and sprawl (the expansion of the built environment into the rural landscape) necessitates discussion of the multi-nefarious impact of road-building into tropical forest. In this case, we can conceive of road building into the rural, tropical, forested landscape as extending urban space and its related problems.

▪▪ Built-environment successes

The built environment also offers animals that are typically generalists and plants that thrive in disturbed environments abundant opportunities to survive and thrive. Food scraps discarded by humans make urban and suburban environments food- and calorie-rich environments. Ornamental plants, scrubs, and trees planted in urban gardens offer seeds and fruit to any animal that can reach them, while many Europeans, Africans, and Asians still maintain allotments—in other words, urban food gardens that provide ample forage for many, many animals. Bears, leopards, foxes, rats, coyotes, monkeys, hyenas, geckos, brown snakes, funnel-web spiders, and, among plants, poppies are good examples of those animals and plants that thrive in the urban environment.

For instance, brown bears in Brasov, Romania, eat garbage from garbage containers at all times of the day and night, both in the periphery of the city and near the old historic city center. They are such frequent visitors to Brasov that a small-scale tourist industry has developed to show tourists Brasov's garbage-eating bears. A number of polar bears are remaining shore-bound in summer and are resorting to consuming food scraps in urban landfills and stray and domestic dogs across the Arctic. Although a very sad example of desperate necessity, this response to the rapid onset of global warming by a number of polar bears is an interesting example of an evolutionary adaptation facilitated by underutilized food resources. In LA, coyotes are common and aggressive! Several coyotes have been filmed approaching young babies, but whether this was out of curiosity or they were looking for food is unknown.

Since 1940, coyotes have expanded their range by 40% and have occupied every state other than Hawaii. Coyotes have expanded their historical western US range into more open, fragmented landscapes, and in northwestern Canada, coyotes appear to have followed the gold rushes of the late 1800s as forest was cut and fragmented. But they have also expanded because of the near-extinction in the lower 48 states of the wolf, and, as noted earlier, the abundance of food in the urban and suburban neighborhoods. This has occurred during a time when in excess of 400,000 coyotes are killed each year in the USA. Across the USA, coyotes have adapted several strategies to survive. Coyote biologist Stan Gehrt has found that in order to avoid human conflicts in urban and suburban USA, these bold, savvy canines have taken to hunting at night, just like foxes and leopards have in Eurasia. Although monogamous territorial animals, Gehrt has also found that when the killing of coyotes escalates, young coyotes mature faster, and females produce large litters, allowing coyotes to quickly reoccupy those territories of killed coyotes. Furthermore, urban and suburban coyotes are healthier and live longer than rural coyotes, for instance, mange and canine distemper are rarely encountered among urban and suburban coyotes. Although coyote numbers continue to grow exponentially across the eastern USA, coyotes have retained an innate fear of humans, and human-coyote conflicts remain few and far between.

Since the mid-1990s, leopard attacks on humans in India, particularly in the urban fringe, have risen as pariah dog numbers have grown. In fact, Sanjay Gandhi National Park that lies adjunct to Mumbai has the highest leopard density in India, with 35 leopards. Many of these leopards patrol the urban fringes and slums, looking for food.

Surveys of leopard scat near Aarey Colony within the NP suggest pariah dogs constitute 40% of the leopard prey biomass while goats only 17%. This 40% represents approximately 1500 pariah dogs per year. Survey author Alexander Braczkowski suggests leopard predation reduces the number of humans bitten by pariah dogs by a 1000 cases and as many as 90 cases of rabies per year in the vicinity of the Aarey Colony. In addition, the leopards save the colony $18,000/year in pariah dog sterilization costs. So, although leopards attack on average seven people across India each year, their hidden benefits far outnumber their human costs.

The growth in leopard population is because Indian vultures have suffered a greater than 97% extinction due to poisoning by diclofenac, a pain reliever and anti-inflammatory drug that was given to sacred cows. An unexpected consequence of the near-extinction of Indian vultures has been the rapid increase in rabies infections and deaths in India, and the massive increase in black kites that have, like the pariah dogs, tried to fill the vulture niche of consuming dead sacred cows at carcass dumps. Today, greater than 6 million kites attempt to fill the void left by the hundreds of thousands of vultures in Delhi. But both the pariah dogs and black kites are not efficient consumers of cow carcasses. Cow carcasses are left to rot, and whereas vultures' stomachs are acidic enough to kill rabid viruses, pariah dog stomachs are not, and so rabies deaths are now 50,000–60,000 individuals a year in India. Rather than persecuting leopards, the Indian government should be educating the Indian public on the benefits of these recent rural-urban migrants. Several other animals are successful colonizers of southern Indian towns; these include the ubiquitous rats, various monkey species, in particular, rhesus macaques, and less so gray langurs and sacred cows.

Red foxes are so common in urban across Eurasian and North America, we forget these are relatively recent exploiters of the abundance of food resources, such as scavenged meat, domestic cats, rats, mice, squirrels, rabbits, birds especially pigeons, invertebrates, and even pond goldfish, that urban environments offer such opportunists. Rats are not a big food item for urban foxes—rather their principle food items are scavenged meat (32.6%), other food items (31.7%), invertebrates (15.5%), wild birds (6.1%), wild mammals (4.9%), fruit (5.3%), and pets (4.0%; mostly cats). Of the other hand, in more industrial environments in the UK, foxes consume more rats and pigeons. Today, black rats (*Rattus rattus*) only persist in the UK due to recurrent introductions in and around ports of entry, industrial centers, and in large urban areas. This suggests the general lack of rats keeps them off the urban fox's menu.

Hyenas once common in downtown Addis Ababa 30 years ago have retreated to the suburban fringe. In the last 30 years, Ethiopia has witnessed a change in diet as their wealth has increased. More people eat meat than ever before. As a result, more animal meat scraps are discarded, but as people have moved out of the downtown, so the spatial distribution of scraps has changed. Today, hyenas' whoops, giggles, and groans are a common nightly chorus in the suburbs, as they have followed the food!

The built environment also offers rock doves, starlings, house sparrows, swallows, martins, American nighthawks, swifts, bats, and peregrine falcons urban cliffs for roosting and nesting, and in the case of peregrines, abundant rock doves to eat. Barn owls are synonymous with barns, hence their name. The vast majority of barn owls live on the urban-fringe nest in barns or in artificial nest boxes. Purple martins across the southern states of the USA are nearly entirely dependent on artificial nest boxes. Rose-ringed and monk parakeets are also expanding across many North American and Eurasian cities after escaping domestic captivity or being deliberately released. Red kites, nearly extirpated in the 1970s from the UK, have expanded and thrived following re-introduction in the 1989 by the RSPB. Today, red kites are frequently observed in suburban neighborhoods and gardens in the southern England consuming food scraps. In the USA, chimney swifts struggle to survive in rural environments if there are no abandoned chimneys or dead or dying trees, whereas chimneys both used and abandoned are common in urban and suburban areas and provide ample nesting opportunities. Humans' predilection to remove all dying and dead trees because of their associated fire-risks means chimney swifts are increasingly more common in the built environment—this is true of the Eurasian swifts where it is unusual to find common swifts outside of the built -environment. Lesser kestrels and white storks across the Mediterranean are also restricted to building nests in the built environment, although storks feed in suburban and rural grasslands and riparian areas. Black redstarts, small, dull-gray flycatchers, are also rare outside the built environment, and in England, they were in the post-WWII era restricted to just old bomb sites in London.

Bird feeders and bird boxes across the UK and the USA (but less so on mainland Europe) in urban and suburban environments provide significant sources of food and shelter for all manner of birds and lots of squirrels too. In the 1960s, back-garden bird feeding started in UK. Coincidently, a few blackcaps had by accident genes that made them fly to the Northwest, not Southwest, in the late fall during migration. Rather surprisingly, these UK wintering birds had higher survival rates than Spanish wintering blackcaps. They also returned to Germany and set up breeding territories earlier. Today, 10% of the German blackcap population winters in the UK. These birds have faced different selection pressures and as a result have shorter more rounded wings and narrower longer bills. The longer bills were selected to eat peanuts through wire-mesh bird feeders. Today, they do not interbreed with the Spanish wintering population. This is an example of Gould's evolutionary contingency, where an unpredictable sequence of antecedent states, not randomness, has led to a situation where the German breeding but UK wintering blackcaps are speciating. This is driven by urban and suburban winter bird feeding.

Blue tits in the 1950s discovered they could tap milk bottles delivered by milkmen and left outside houses—they pierced the foil tops and drank the cream that had risen to the top of the milk bottles. Eurasian collared doves, once restricted to Eurasia east of Austria, exploded across Europe in the 1950s. This was followed by their expansion across the USA following their accidental introduction to the USA via the Bahamas in 1974. Today, Eurasian collared doves inhibit every US state. Prior to expanding across western Europe and North America, Eurasian collared doves are thought to have originated in South Asia spreading to the Balkans by the late 1500s. In the USA, they colonized urban areas where food is abundant, especially grain stores and bird feeders, and then moved on into the rural landscape. The expansion of the Eurasian collared doves has been even more rapid than the expansion of either house sparrows or starlings. In the case of house sparrows, repeated introduction to states was facilitated by Eugene Schieffilin after the initial release of eight pairs released in Brooklyn, New York, in 1851. European starlings, on the other hand, have had no further assistance since the release of 60 birds by Eugene Schieffilin in 1890 in New York City's Central Park. Today, some 200 million reside in the USA. Both species are now found in the urban and suburban environments throughout Central and South America and the Caribbean, making them among the most successful urban colonists of all time.

However, even the Eurasian collared dove, house sparrow, and European starling pale in comparison to the expansion into and colonization of human dwellings of both brown and black rats, cockroaches, and recently bedbugs. At least 78% of US homes are home to cockroaches. Although cockroaches are generally harmless to humans, their feces, salvia, and body parts contain certain proteins that trigger allergic reactions or asthma among children. Approximately 7 million US children suffer asthma that is incurable, yet removing cockroach-derived allergens does limit asthma attacks. Cockroaches can also spread *E coli*, salmonella, *staphylococcus aureus,* and streptococcus, several parasitic worms, and act as vectors for typhoid fever, cholera, amoebic dysentery, leprosy, plague, campylobacteriosis, listeriosis, giardia, and the polio virus.

Rats are known to act as vectors for 35 human pathogens. Rats can transmit these pathogens through their urine and feces (e.g., hantavirus, hemorrhagic fever with renal syndrome, lassa fever, South American Arenaviruses), or saliva through bites (e.g., rabies, rat-bite fever), or through their fleas (e.g., plague). Indirectly, rats can also transfer to humans cutaneous leishmasiasis, lyme disease, and West Nile fever, among many others. Rats are such effective vectors of human pathogens because they have the same basic physiology, similar organs and body parts. Rats are very hardy, very intelligent, very adaptable, and habitat generalists—so, in other words, perfectly adapted to life among the built environment. At least 2 million rats call NYC home, although this is generally believed to be an underestimate. According to the CDC, in order to combat urban rats, we need to seal up holes inside and outside the home to prevent entry, trap rodents within and around homes, and clean up rodent food sources and nesting sites. Just as Russia and Alaska have bear-proof garbage bins, so rat-proof garbage bins are also necessary to reduce rat populations. Bedbugs are the new wave of colonizers! Today, most bedbugs are immune to insecticides, although not immune to washing clothes in greater than 160 °F water. Evolutionary selection pressures since the mid-1960s has meant that although each new bed-linen pesticide would kill more than 99% of bedbugs exposed, natural bedbug genetic variation would mean a few individuals would survive each new pesticide. These survivors colonize bed-linen across the world aided by our human proclivity to travel. An evolutionary arms-race has ensued, with the bedbugs looking to have recently won. Thankfully for the human race, bedbugs, although annoying, generally do not act as vectors for human pathogens.

Nevertheless, University of Toronto researchers Marc Johnson and Jason Munshi-Smith suggest that human-environmental interactions within the built environment drive contemporary evolution today, bedbugs being a classic example. Their work suggests that those species (and the diseases they vector) that can adapt and thrive in the built environment will have a significant impact on human health into and beyond the twenty-first century. The built environment will also change the evolutionary directions of many, if not all, of those species that adapt to living in the built environment. For instance, white-footed mice have genetically differentiated at the scale of city parks in NYC. In other words, unique populations of white-footed mice that are genetically different exist in each major NYC park. In London, mosquitos in the London Underground have evolved to live without the need for blood to produce eggs, and they also remain active throughout the year.

8.3 Urban Heat Islands

> Urban heat islands (UHIs) are areas that are significantly warmer than surrounding rural areas due to human activities and the built environment. Heat islands are primarily the product of the reduced albedo of urban surfaces, in other words, roads, runways, pavements, buildings, and roofs are darker than natural vegetation and absorb more short-wave radiation than natural vegetation and then emit more long-wave radiation that heats the adjacent air.

The built environment, especially those areas in and around the central business district (CBD), also produces more heat in the form of waste heat generated by energy use, for instance, hot car and truck exhausts, air conditioning exhausts, preparing food at home and commercially, manufacturing exhausts, and through electrical generation. Areas downwind of UHIs frequently suffer increased rain events, for example, northeast, east, and southeast Atlanta suffer more thunderstorms than areas to the west of the central business of Atlanta. Although the advection of air masses due to excessive heat over Atlanta certainly contributes to thunderstorm anomalies, so do the elevated volumes of concrete dust. Simply put, fine airborne dust particles generated by elevated volumes of road traffic on concrete roads increase the number of condensation nuclei above urban areas.

The heat islands and their associated rain/thunderstorm anomalies are not spatially contiguous. Heat islands are at their most intense in the CBD and immediate surrounds, while thunderstorm anomalies are located downwind in the suburban environments. Their differential human impacts are significant. Across the USA and beyond, heat stress can cause illness and death. Neighborhoods in and around the CBD typically exhibit fewer socioeconomic opportunities and larger concentrations of ethnic minorities, and suffer from higher urban densities, little green space, fewer shade trees, and a higher proportion of poorly maintained rented domestic buildings. As a result, these communities have less resources to cope with the higher heat stresses. As a consequence, Phoenix and Atlanta have disproportionately higher mortalities in summer than surrounding suburban areas. However, hail and wind damage and flash flooding impact more suburban neighborhoods than CBDs. However, these neighborhoods have the material resources to cope with these weather anomalies.

A third telling example is that of Dhaka, Bangladesh, which, according to the United Nations, is expected to crest the 21 million resident mark in the year 2020, doubling its population from the year 2000. Much of that population growth was accompanied by construction and physical expansion of the city, such that the largest land use change over two decades has been from agricultural to residential or commercial. This large-scale and rapid land use change leads to a significant increase in the urban heat island effect. As one way of combating the UHI effect, Bangladesh's Department of Agricultural Extension began a program in 2013 to provide rooftop gardening training to some of Dhaka's residents. Greening rooftops decreases the amount of solar radiation absorbed by buildings, increases evapotranspiration (thus cooling), and can provide additional shade. Moreover, recently, the city of Dhaka is considering providing tax breaks for residents who plant rooftop gardens, which not only helps cool the building in the summer but also provides added insulation in winter months. The hope is that this incentive will add significantly to the current 6000 rooftop gardens in Dhaka, which provide numerous benefits in addition to UHI mitigation. For instance, rooftop gardens or greenspaces help decrease stormwater runoff, provide cleaner air, and hold great potential for recreation and urban agricultural production. Studies have shown that rooftop gardens retain over 50% of stormwater runoff (on average) and up to 90%, depending upon the type of garden and storm event.

Indeed, many cities in the global north have promoted rooftop gardens through legislation and/or monetary incentives. Europe has long taken advantage of the benefits of adding greenspace to urban rooftops, while Germany has mandated and supported greening rooftops in urban areas for decades. Recently, France passed legislation in 2016 requiring some green-roofing or solar panels on new commercial construction. Although it is much less common in North America, around 25 cities in the USA and Canada have some type of green roof policy or incentive. These range from Toronto's rooftop legislation from 2009 requiring all new industrial buildings to include some rooftop greening, to the requirement in San Francisco that most new construction contain at least 15% green or solar roofing, to New York City where they provide tax breaks and have embarked on an effort to plant a million trees above ground.

Even roads in rural areas generate both warmer micro-climates or linear heat islands and more abundant food resources through road-kill that allow species to penetrate into areas and regions where once they were rare or nonexistent. Tarmac roads absorb more incoming solar radiation, retain this energy longer, and emit this for a longer period into the night than the surrounding vegetation. Tarmac roads also continue to absorb incoming solar radiation into the fall and winter, and accelerate snow melt.

8.4 The Built Environment and Weather Hazards

In New Orleans, the impact of Hurricane Katrina in 2005 was more socioeconomic than physical. Neighborhoods of fewer socioeconomic opportunities were hit much harder and suffered higher mortalities than richer neighborhoods, even though the whole city was hit by Katrina.

In Tornado and Dixie Alleys, residents of mobile homes are ten times or more likely to die in deadly tornadoes than in a permanent home. In 2000–2012, the five states with greatest number of tornado-caused deaths were Alabama (293), Missouri (233, including 158 in the 2011 "Joplin Tornado"), Tennessee (132), Georgia (62), and Mississippi (61), and most people were killed in mobile homes (426), followed by permanent homes (422; half of those in 2011), vehicles (79), businesses (124), and in the open (27). For example, all but nine of the 6600 mobile homes in the path of Hurricane Andrew's traverse across South Florida were destroyed in 1992. Rather troublingly, Mississippi and Alabama rank third and fourth in the nation in the percentage of people living in mobile homes, and these people exhibit high levels of poverty and functional literacy. These attributes partially explain why 61% of fatalities in mobile homes occur at night, where low levels of public education and literacy and a lack of hazard sign training contribute to the misunderstanding of hazard signs and warnings.

8.5 Climate Adaptation in Cities Worldwide

Approximately three-fourths of the carbon dioxide emissions from energy use worldwide come from cities, and given global urbanization projections, this trend will only increase in the coming decades. Thus, in many ways, cities are, and will continue to be, on the front lines when it comes to dealing with the myriad effects of a changing climate. Indeed, the rapid pace of urbanization in many parts of the Global South and the projected increases in urbanization, as citizens migrate from rural to urban areas, means cities will continue to consider climate change resilience in the coming years. From transportation networks, to wastewater and sewage infrastructure, flood risk, saltwater intrusion, and potable water access, cities worldwide will be forced to deal with the effects of climate change (if they are not already having to do so). In addition, the infrastructure improvements these growing cities need to accommodate new residents will only add more carbon dioxide into the atmosphere. The fact that cities tend to house large populations exacerbates these challenges, although relatively high population densities of many urban areas may help with planning for alternative transportation (i.e., non-automobile options). Low-carbon construction practices coupled with increased urban planning and design must be a priority for such fast-growing urban areas. The rub, of course, is how to pay for such infrastructure improvements and climate change adaptation strategies.

For instance, although there remain concerns about scientific data accuracy and availability, particularly in Latin America, Africa, and Asia, most cities worldwide rank obtaining the necessary resources to meet the planning and implementation goals as much more difficult and thus their primary challenge. Indeed, the challenge in obtaining the financial resources needed to plan and implement strategies to increase climate resiliency of urban infrastructure is universal among cities worldwide. Given this dearth of financial resources, most cities are working to incorporate climate considerations into existing planning and/or disaster management strategies. In addition, many have turned to creating policies and communication strategies as opposed to focusing on the much more expensive task of improving the climate resilience of infrastructure.

Despite the similarities in most cities' experiences, regional differences do exist regarding the types of challenges cities have encountered when pursuing climate change adaptation plans. For instance, urban areas in the USA have relatively low levels of engagement in climate adaptation preparation, when measured as the share of cities actively working on climate adaptation, compared to cities in other regions (e.g., Latin America, Europe, Africa, and Asia). Moreover, most cities that are working in this area are in the early preparation stage (and not yet in the planning or implementation stages). Furthermore, US cities suffer from a general lack of substantial support from both national and local governments. By contrast, European, Asian, and Latin American cities report significant support from national governments, yet still characterize support by local governments and officials as inadequate. Perhaps not surprising, NGO involvement is more prevalent in Africa than in other regions. Nevertheless, there are myriad opportunities for reducing the carbon footprint of cities by promoting the creation of green spaces, vegetation corridors, urban forests, and the like, which will help to offset the carbon outputs from energy production and transportation while simultaneously curbing flood risk by soaking up water. Ideally, all of these climate change mitigation strategies would be adopted in both Global South and Global North cities along with a commensurate commitment from urban residents to alter their consumption practices so as to decrease the overall GHG emissions originating from the world's urban areas.

Informal Urban Settlements Informal urban settlements including slums, shantytowns, and squatter settlements are growing rapidly, especially in the Global South, as cities struggle to keep up with the pace of rural-to-urban migration. Estimates predict that by 2050, three billion people may be living in these informal urban settlements that often lack basic infrastructure such as adequate housing, electricity, potable water, and sewage systems. In addition to the challenge of providing sufficient infrastructure to these growing residential areas, city and national governments must consider how to increase climate change resilience in these parts of their cities. Given their relatively high vulnerability of informal urban settlements, climate scientists, geographers, urban planners, and the state at multiple geographic scales should focus on prioritizing these neighborhoods and their residents in climate change adaptation planning. Fortunately, they may not need to start from scratch in many Global South cities, as local, grassroots efforts at adaptation have been underway for years or even decades. These efforts supported by local groups, NGOs, and/or local municipalities have worked to map flood-prone areas for instance, and these projects may serve as models for other cities in the Global South. In many ways, this makes much more sense than attempting to transplant models from the Global North, where physical, urban, and population geographies frequently are drastically different than the situation in Global South cities, which are much more likely to contain one or more informal urban settlements. So if these locally produced plans in Global South cities can be scaled up, their tools and models may be transferable to other Global South cities.

Environmental Justice/Racism As Robert Bullard states in the latest edition of his seminal text *Dumping in Dixie: Race, Class, and Environmental Quality*, "To be poor, working-class, or a person of color in the United States often means bearing a disproportionate share of the country's environmental problems." Bullard, who first wrote *Dumping in Dixie* in 1990, is widely considered a pioneer of environmental justice research and activism. The environmental justice movement traces its roots to the small town of Warren, North Carolina, USA, in 1982, when United Church of Christ members turned to their church to provide assistance in fighting the siting of a polychlorinated biphenyl (PCB) waste dump in their neighborhood. Ultimately, the church members lost that battle since the waste dump was built there, but from that initiative came the United Church of Christ's Commission on Racial Justice study published in 1987. The study, titled *Toxic Wastes and Race in the United State*, analyzed the geographic distribution of waste disposal sites, polluting industrial facilities, and hazardous waste facilities across the USA. More specifically, the report examined the relationship between where such facilities were located and race, ethnicity, and socioeconomic status, and race was found to be the key predictor of the geographic location of these facilities. Indeed, socioeconomic status also was significant, but not as significant as the racial and ethnic composition of the neighborhoods in which these environmental and human health hazards were located. The release of toxic wastes demonstrated for the first time what many people of color in America knew in 1987—that they were much more likely to suffer health impacts from polluting facilities than were whites and that it was not merely a coincidence. Thus, the environmental justice movement was born.

Today, approximately 80% of hazardous waste industrial polluting facilities in the USA are located in urban areas. The geographic patterns of these environmentally degraded/degrading locations are not coincidental; they are a symptom of structural inequities inherent in social and economic systems around the globe. Environmental justice advocates point to the disproportionate number of toxic waste dumps in disadvantaged neighborhoods in the Global North and Global South, while also shining a light on inter-regional examples like the shipping of electronic waste from the Global North to the Global South. Much of the recent research in environmental justice, moreover, has moved beyond a focus on distributional justice (i.e., where the environmental "bads" and "goods" are sited in a given area and thus who is impacted negatively or positively) to a broader and more sophisticated view of justice. For instance, studies now approach environmental justice from procedural justice perspective, focusing on democratic (or lack thereof) decision-making processes that lead to disproportionate impacts on communities of color. Other researchers consider the capabilities needed to create or restore healthy, functioning communities (i.e., justice of capabilities). Moreover, environmental justice researchers have broadened their research focus to include work on the effects of natural disasters, uneven impacts of climate change, and urban greenspaces.

8.6 Brownfields

Definition

A brownfield is a previously used property wherein pollutants, hazardous substances, and/or contaminants are (likely) present. Many brownfields are former industrial or commercial properties that have long been abandoned, but still (likely) contain some amount of pollutants or toxic agents contaminating the soil.

According to the US Environmental Protection Agency, there are approximately 450,000 brownfields in the United States today, and many of these sites are located in urban areas. The presence of a brownfield site holds myriad health and safety implications for the surrounding communities and oftentimes adversely impacts the chances of clean-up, remediation, and/or redevelopment of the land. Redeveloping brownfields not only remediates the contaminated land and/or water, it also emits fewer greenhouse gases than new (i.e., "greenfield") development, by conserving existing forested land.

Of course, the evaluation and measurement of a successful brownfield redevelopment project both depends on your goals and on resources available. Despite laudable community goals of economic development (e.g., job creation and augmenting the local tax base), improving public health, environmental sustainability, or enhancing a sense of neighborhood, most projects must be profitable to generate the necessary private investment. Moreover, many community-based efforts are time-consuming, expensive, and are not necessarily the main priority for local governments and potential private investors. In short, although it is not *only* about the money, it is still about the money.

Costs (both known and unknown) may be prohibitive to the extent that cleanup and redevelopment efforts far exceed any potential monetary gain from new development. In these (all too common) scenarios, brownfield sites may sit empty for decades, not only aesthetically displeasing but also potentially dangerous to surrounding communities. Nevertheless, there are some brownfield redevelopment success stories such as Atlantic Station in Atlanta, GA. Atlantic Station is a multi-use development located in midtown Atlanta on the 138-acre site of the old Atlantic Steel Mill. It opened in 2005 and has over 2.5 million square feet of office, retail, and residential space, including IKEA and Target as anchor stores. Recently, Atlantic Station has undergone even more construction, adding tens of thousands of square footage to its office and residential space and redesigning some of its greenspaces as well.

Sustainability and Cities (Fig. 8.3)

Masdar City in Abu Dhabi, part of the oil-rich United Arab Emirates (UAE), which bills itself as "one of the world's most sustainable urban communities," is a lesson in both the technological complexity and costs associated with creating (or re-creating) truly environmentally sustainable cities. More than 500 companies from nearly every continent have operations in Masdar City, yet despite projections of 50,000 residents, it is currently home to a mere 1300 (and most of those students in university housing of the Khalifa University of Science and Technology). This has led some observers to wonder whether this experiment at urban sustainability has instead become a ghost town. Others are even more critical of such ventures, arguing that such eco-utopias are representations of a singular focus on technological fixes to the world's impending environmental crises, with little to no thought of the changes required of consumption practices of us in the Global North, much less the plight of those most vulnerable to the impacts of climate change.

A synopsis of Abu Dhabi's descriptions of Masdar City over its short lifespan highlights the realities of implementing environmentally sustainable urban practices on a large scale, even for a hugely wealthy country like the United Arab Emirates. First, in 2007, the city was billed as an "eco-city" before being labeled "zero carbon" city. Soon thereafter, it was dubbed a "zero carbon" city until officials realized this was impossible and hence quickly labeled it a "low carbon" city. In 2015, UAE officials settled on "smart city" to adequately describe its urban project, and most recently the preferred moniker appears to be "one of the world's most sustainable urban communities" according to its official website. Critics of the project point to these multiple examples of an identity crisis over its short lifespan as evidence that the "eco" might better reflect Masdar City as first and foremost an economic venture rather than an ecological one.

Moreover, despite the city's use of sustainable materials, solar power, and sustainable urban design elements, critics have drawn attention to issues that appear when we look just below the surface. For example, the city's construction relied heavily on imported resources, one example of which is coltan, a mineral found in

Fig. 8.3 Original smokestack from old Atlantic Steel Mill at Atlantic Station (Atlanta, GA), with newly constructed building in the background. (@ Richie Yarbrough 2020)

lithium ion batteries and many smart technology devices. The vast majority of the world's coltan is mined in Democratic Republic of Congo where numerous international organizations, including the United Nations and Amnesty International, have identified exploitative working conditions, myriad human rights violations, and environmental degradation. Add to that the documented subpar working conditions of the primarily South and Southeast Asian labor force that has built Masdar City, and a quite divergent picture emerges of the sustainability of such an urban experiment. This all-too-common experience of migrant workers in the construction of such eco-utopias like Masdar City ends with laborers leaving for the next project, oftentimes in another city or country that may or may not afford them rights as workers, healthcare, and fair pay.

Many geographers and other researchers argue that as we continue to examine green technological innovations that might aid cities in mitigating and/or adapting to climate change, research must also interrogate the experiences of the workers on the backs of which these solutions are built. In addition, any strategies or plans to mitigate climate change or enhance climate resilience must account for and seek to address the uneven geographic patterns of implementation across and within countries and in both urban and rural contexts. To put it more concisely, climate change solutions cannot ignore, or worse exacerbate, existing divides between the world's haves and have-nots, leading to a variegated geography of gated eco-bubbles. Otherwise, we cannot claim any true commitment to urban environmental justice.

Summary

Today, a majority of the earth's residents live in cities. In the coming decades, this number is expected to increase such that, by 2050, projections expect two-thirds of the world's residents will reside in cities. Urbanization rates are highest in Global South countries, as millions of developing countries' residents arrive in cities from rural areas each day. These urbanization trends mean that by 2050, the world will see 90% of the growth in urbanized populations concentrated on two continents, Asia and Africa. The implications for impending global environmental issues are myriad and somewhat dire, as cities will continue to be on the front lines of the global fight to keep the climate crisis at bay. We are already seeing cities across the world plan for the realities of sea-level rise and increased energy production/consumption, while Global South cities also work to tackle their current and future infrastructure and housing needs. Notwithstanding the recognition of the extant and future challenges cities face, there remains a dearth of resources across all geographies to address the coming climate crisis. This is particularly the case in Global South cities, where financial resources are even more scarce than in many Global North contexts.

Moreover, far from being pushed out of sprawling urban areas, many examples of nature "fighting back" abound, from leopards in the streets of Mumbai, to hyenas in the expanding suburbs of Addis Ababa, Ethiopia, to a multitude of bird species occupying urban and peri-urban landscapes across Europe. Although the future growth of cities across the world is not in doubt, significant questions remain regarding what should and can be done to plan for urban heat island effects, sea-level rise, and related climate change impacts. One thing is for sure, however. Global South cities cannot follow the consumption and sprawling urban development path of many Global North cities if the cities across the world are to meet the environmental challenges of the twenty-first century, while honoring a moral commitment to assist and protect the most vulnerable residents of our shared planet. Given the uneven geographies of the causes of climate change, residents of the Global North have an obligation to lead in combating climate change. Technological advancements will indeed play a role, but the shared sacrifices required of urban residents across the globe must weigh heaviest on those Global North urbanites, whose disproportional contribution to the climate crisis is continuing unabated.

Further Readings

Bai, X., et al. (2018). Six research priorities for cities and climate change. *Nature, 555*(7694), 23–25.

Barrington-Leigha, C., & Millard-Ballb, A. (2015). A century of sprawl in the United States. *Proceedings of the National Academy of Sciences., 112*(27), 8244–8249.

Braczkowski, A. R., O'Bryan, C. J., Stringer, M. J., Watson, J. E., Possingham, H. P., & Beyer, H. L. (2018). Leopards provide public health benefits in Mumbai, India. *Frontiers in Ecology and the Environment, 16*(3), 176–182.

Cugurullo, F., & Ponzini, D. (2018). The transnational smart city as urban ecomodernisation: The case of Masdar City in Abu Dhabi. In A. Karvonen, F. Cugurullo, & F. Caprotti (Eds.), *Inside smart cities: Place, politics and urban innovation*. London: Routledge.

Gehrt, S. D., & McGraw, M. (2007). Ecology of coyotes in urban landscapes. In *Wildlife Damage Management Conferences–Proceedings* Ohio State University. Columbus, OH, USA. (p. 63).

Johnson, M. T., & Munshi-South, J. (2017). Evolution of life in urban environments. *Science, 358*(6363), eaam8327. https://doi.org/10.1126/science.aam8327.

McArdle, M. (2018, April 24). Is Masdar City a Green Lab or a Ghost Town? *Popular Science*.. https://www.popsci.com/masdar-city-ghost-town-or-green-lab/

Nicholls, E., et al. (2018). Monitoring neonicotinoid exposure for bees in rural and peri-urban areas of the UK during the transition from pre- to post moratorium. *Environmental Science & Technology*. https://doi.org/10.1021/acs.est.7b06573

O'Neill, D. (2019). Survival by degrees: 389 bird species on the brink.. Audubon. NAS. Audubon's Climate Science Series Reports.

United Nations, Department of Economic and Social Affairs, Population Division. (2018). *World urbanization prospects: The 2018 revision*. Online Edition.

Practical Solutions

Contents

M. R. Welford, R. A. Yarbrough, *Human-Environment Interactions*,
https://doi.org/10.1007/978-3-030-56032-4_9

Learning Goals

After reading this chapter, you will be able to:

- Evaluate whether there are sustainable approaches to resource use, organic farming, or non-fossil fuel-based transportation systems that can be implemented in all geographic contexts.
- Explain whether ecotourism is primarily a mechanism for environmental conservation or economic development.
- Articulate the implications of the commodification of environmental resources for the ways in which environmental policies are debated and implemented.
- Assess whether political sovereignty should supersede global environmental concerns.
- Evaluate what kind of environmental future might be on the horizon and where.
- Argue whether we as a planet and species have a future.

9.1 Are There Practical Solutions to the Crises We Face?

9

Are we looking at an abyss? As Greta Thunberg so succinctly stated and with such passion and conviction in her address to the UN's Climate Action Summit in NYC on September 23, 2019:

> People are suffering. People are dying. Entire ecosystems are collapsing. We are in the beginning of a mass extinction, and all you (*the world's leaders*) can talk about is money and fairy tales of eternal economic growth. How dare you! Greta Thunberg, 2019, U.N.

Here Greta is echoing Joseph Stiglitz's (and others) views of our global economy. Stiglitz argues that Thatcher and Reagan destroyed the social contract between workers and executives by creating an upside-down global economy that:

> created a society in which materialism overwhelms moral commitment, in which the rapid growth that we have achieved is not sustainable environmentally or socially, in which we do not act together to address our common needs. Market fundamentalism has eroded any sense of community and has led to rampant exploitation of unwary and unprotected individuals. Joseph Stiglitz, 2010.

The *moral depravity* of the financial operators liberated by Thatcher and Reagan in the 1980s, and more recently under Yeltsin and Putin in Russia, Rao and Singh in India, and Deng-era reforms in China, has allowed the global financial sector to exploit poor and middle class and move money from the bottom toward the top earners at rates not seen since the railway magnets of the late 1890s in the USA.

Greta then returned to global environmental change and reiterated the UN's own Intergovernmental Panel on Climate Change and recast their prognosis:

> The popular idea of cutting our emissions in half in 10 years only gives us a 50% chance of staying below 1.5 degrees [Celsius], and the risk of setting off irreversible chain reactions beyond human control.

> Fifty percent may be acceptable to you. But those numbers do not include tipping points, most feedback loops, additional warming hidden by toxic air pollution or the aspects of equity and climate justice. They also rely on my generation sucking hundreds of billions of tons of your CO2 out of the air with technologies that barely exist.
>
> To have a 67% chance of staying below a 1.5 degrees global temperature rise – the best odds given by the [Intergovernmental Panel on Climate Change] – the world had 420 gigatons of CO2 left to emit back on Jan. 1st, 2018. Today that figure is already down to less than 350 gigatons.
>
> How dare you pretend that this can be solved with just 'business as usual' and some technical solutions? With today's emissions levels, that remaining CO2 budget will be entirely gone within less than 8 1/2 years. Greta Thunberg, 2019, U.N.

What can we take from this speech? Firstly, Greta is absolutely correct—we cannot wait for technological solutions that have not been developed; we cannot act like "business as usual," something *The Limits to Growth* argued against in 1972. We also cannot hide our heads in the sand and hope these global environmental issues go away, or that they will only hurt our future generations, or that some supernatural power will solve these problems, or that we can move en masse to some other new planet (*there is no Planet B*), or be raptured—we must, must act now at a personal level, at the levels of local, regional, national, and supra-national governments, and we must act together with a deep understanding that we are embroiled in a global "Tragedy of the Commons"!

9.2 Contingency, Resilience, Recovery, and Response

Irrespective of one's political viewpoint, the long-term solution to our current environmental/resource/pollution crisis is to have fewer people on earth and for those to consume less. However, birth control is a cultural/political hot topic and unlikely to be addressed. Nevertheless, it is completely fair to state that the developed world has a resource-consumption and a pollution-production problem. Furthermore, it is not fair for the developed world, having already exploited our own environment, to ask others not to exploit theirs unless we can bring our own rate of consumption under control. For instance, the USA, with less than 5% of the world's population, uses 24% of energy and produces 19% of the world's garbage.

At local and regional scales, conservation remains problematic. Money breeds corruption, and as soon as money is available to protect endangered species, people who were not otherwise interested in species become interested in the money. This can lead to conservation organizations, whose budgets go mostly to maintaining central offices and staff with a pleasant lifestyle rather than to projects that save species in the field. Human motivations are often mixed and muddled, and species suffer such as the Hawaiian crow. Furthermore, competent people are in short supply; endangered species may have their fates determined by people who may be well-meaning but unskilled or incompetent. And let us not forget, most scientists do not make effective lobbyists—if they know of an alternative interpretation, they feel obliged to mention it. Their honesty and objectivity may confuse the public and make a clear choice difficult.

Some extinctions take with them a piece of human history and culture. Some people believe all species have rights, including the right to exist: we do all share our planet. We humans happen to have evolved language and technology and to be in control at present, but that does not mean we have an exclusive or "natural" right to dominate. We do have the responsibility to control ourselves. Some things are better left alone—it is easy to do too much too quickly without thinking about alternatives or consequences. Sometimes, constant management is the only way to preserve some species—gorillas of Congo, or the kestrels of Mauritius, or whooping cranes of the USA and Canada.

But what are our priorities? According to an

> anthropocentric view, the environment has value only for what it can provide for us.

Others suggest an instrumental view, which holds that the environment has values because it helps people to reach some end such as obtaining food, shelter, clothing, medicine, or entertainment in the form of ecotourism or hunting. If we respect the.

> ecosystem services, ecosystems processes that we often take for granted include such things as clean water, timber, and habitat for fisheries, and pollination of native and agricultural plants.

A more exhaustive list include services that moderate weather extremes and their impacts; disperse seeds; mitigate drought and floods; protect people from the sun's harmful ultraviolet rays (e.g., ozone); cycle and move nutrients; protect stream and river channels (trees, beaver dams) and coastal shores from erosion; detoxify and decompose wastes; control agricultural pests (controls include birds, bats, spiders, and hedges); maintain biodiversity (forests); generate and preserve soils and renew their fertility; contribute to climate stability; purify the air and water; regulate disease-carrying organisms (vultures); and pollinate crops and natural vegetation (bees, birds, bats, etc.). Constanza and colleagues suggest ecological services across the world and our biosphere have a value of US$18 trillion per year.

Nevertheless, the frontier ethic assumes that the earth has an unlimited supply of resources. If resources run out in one area, more can be found elsewhere (e.g., colonial system), or alternatively, human ingenuity will find substitutes (e.g., whale oil to crude oil). This attitude sees humans as masters who manage the planet. This mirrors the Judeo-Christian or Dominion Ethic that suggests the following: "And God blessed them, and God said unto them, Be fruitful and multiply, and fill the earth and subdue it; and have dominion over the fish of the sea and over the birds of the air and over every living thing that moves upon the earth." Deviating from these two ethics is the Stewardship Ethic that claims we, as humans, have superior intellect, so it is ethically correct that humans act as stewards of the land. Both of these demonstrate an anthropocentric ethic.

In contrast, there is the

> biocentric ethic, which argues that all life possesses intrinsic value, and as such, the environment has inherent value just because it exists. An individualistic biocentric ethic recognizes intrinsic value in every living thing. A holistic biocentric ethic recognizes species or aggregates of living things; however, some argue that it is not possible to have a holistic approach because "species" are not living! An egalitarian biocentric ethic accords equal value to all living things (e.g., Jains), whereas, a non-egalitarian biocentric would give greater value to certain living things over others (possibly because of their ecosystem worth).

Taking this logically to conclusion, an

> ecocentric ethic argues that all aspects of the environment, both living and nonliving, have inherent value.

The battle between Pinchot (chief of the US Forest Service) and Conservation and Muir (future leader of the Sierra Club) and Preservation in the US in the 1920s–1930s led Aldo Leopold to consider a more inclusive land ethic. His proposal, written up in the Sand County Almanac, 1939, heralded the beginning of environmentalism. Leopold thought that ethics direct individuals to cooperate with each other for the mutual benefit of all. He argued that this "community" should be enlarged to include nonhuman elements such as soils, water, plants, and animals, "or collectively: the land." In the USA, this work was followed with Rachel Carson's 1962 Silent Spring that launched the contemporary American Environmental Movement and Garret Hardin's 1968 Tragedy of the Commons.

In 1975, Singer revived environmentalism with his advocacy for freeing all animals from use by humans, whether for food, medical testing, industry, personal adornment, entertainment, or anything else. This has resonated strongly among many people including vegans, those opposed to Sea World, those opposed to circuses, those opposed to medical testing (Peta). However, Singer did not advocate for equal treatment of all animals. At around the same time and motivated by Singer, animal geographies developed. The earliest animal geography focused on cataloging wild species, their spatial distributions, and their environmental adaptations. A second wave of animal geography studies domesticated animals and the ways in which livestock are and were enmeshed in human cultures, as well as the impact of livestock on the landscape. The third wave of animal geography emerged in the 1990s, with the rising visibility of animal-based social movements, new scientific understandings of animals as well as human impact on biodiversity, and new developments in social theory embedded within animal geographies. It focused on the full socioeconomic spectrum of human–animal relations including wild and farmed animals, as well as pets, captive animals, research animals, and entertainment animals—and especially the ways animals are used as cultural signifiers.

At around the same as Singer's work, deep ecology, developed by Arne Naes, emerged.

> Deep ecology recognizes the inherent worth of all other things, aside from their utility. The philosophy emphasizes the interdependence of organisms within ecosystems (of which we are a part of) and that of ecosystems with each other within the biosphere.

Clearly, if we are to follow deep ecology, we have to prioritize conserving and preserving all things and all animals. But given that 83% of all wild animals have been exterminated, where should our priorities lie? Should we prioritize conserving and preserving those animals, processes, and elements of the biosphere that offer the greatest positive and healthy impact on ecosystems processes? Do we accept that the black-breasted puffleg, a beautiful, range-restricted hummingbird species found on just two volcanoes in Ecuador, is beyond help! Do we look at what can be done at different scales—those at individual, community, and national legislative levels? This is in fact what we choose to do.

9.3 What Can *Individuals* Do to Mitigate Climate Change, the Sixth Mass Extinction, and Natural Resource Exhaustion?

We must explore ideas on how to mitigate and prepare for the challenges that face humanity—a change in the dominant energy source, food availability and scarcity, climate changes and resulting impacts on affected societies. We feel that these ideas go beyond a "we must" and rather provide a blueprint of sorts for a transition to a more sustainable future; clearly, our standard of living and expectations are running on borrowed time.

There are many actions we can take as individuals to reduce our global spatial, resource consumption, and pollution footprint!

▶ Example

Coffee is an excellent example of what can be obtained if consumers change their consumption patterns. Today, the organic coffee market is much more than a niche market—it represents a third of all organic beverages sold worldwide at a value of US$50 billion. However, multinational, agribusinesses that have embraced organic production have also contributed to the recent acceleration in Amazon deforestation as these businesses seek rapid implementation of organic crops in particular organic soybeans.

According to Amanda Rodewald and coauthors, moving to shade-grown coffee is cost-effective and more profitable than sun-tolerant, high resource-dependent (lots of pesticides, herbicides, and fertilizer) coffee. Although shade-tolerant coffee scrubs produce less coffee per scrub, they cost less to maintain, as forests provide free insecticides, birds, more moisture (trees provide more shade, reducing soil evaporation), and nutrients through leaf-fall—in other words, the trees providing shade provide free ecosystem services. Shade-tolerant coffee plantations also require less labor. Mixed-species bird flocks that occur in tropical coffee-producing areas and include wintering neotropical migrants from the NA can, according to Rodewald, save 23–65 pounds of coffee through their very efficient and directed insect consumption. ◀

Decreasing Your Carbon Footprint

But what are the most effective actions individuals can take to reduce their carbon footprint? Seth Wynes and Kimberly Nicholas recommend four high-impact actions (those actions that result in low emissions) that would substantially reduce individual annual carbon emissions:

1. Have one fewer children
2. Live car-free
3. Avoid plane travel
4. Eat a plant-based diet

According to Wynes and Nicholas, in the Global North, 58.6 tons CO2-equivalent (tCO2e) would be saved each year if you had one fewer child. If you went car-free, you would save 2.4 tCO2e per year. If you ceased to fly, you would save 1.6 tCO2e per transatlantic crossing. If you ate a plant-based diet, you would save 0.8 tCO2e each year. Rather surprisingly, more conventional approaches to reducing individual carbon footprints cannot match these big four individual choices. For instance, Wynes and Nicholas found that recycling (something we should still do) was four times less effective at reducing individual carbon footprints than eating a plant-based diet. Changing light bulbs to energy-efficient bulbs was eight times less effective than eating a plant-based diet.

Even if individuals are keen to adopt high-impact actions to reduce their carbon-emissions, local, regional, and national structural and cultural impediments might impede individual actions. For instance, living in sprawling cities like Bogota, Atlanta, and Cape Town promote high carbon usage, while cultural traits such as a Western meat-based cuisines and cultural expectations to conform to these dietary norms make transitions to a plant-free diet difficult, especially when eating out! Elsewhere, individuals are flaunting cultural norms; in Eastern Europe, Italy, China and Japan, more women are delaying having children, seeking higher education instead and fulfilling career paths. Others are having no children or fewer children for social and economic reasons, deterred by the high cost of living and of raising children, job instability, and especially in China a lack of parental leave.

Sadly, the EU, the USA, Canada, and Australia encourage schools to promote low-impact actions. Let us look in more depth at these high-impact actions, and then at lower-impact actions, that are still necessary.

High-impact actions that significantly reduce individual carbon emissions.

If you are single or married but childless, do not have any kids! Fewer Global North children means less plastic toys and less clothing being bought for each child, less food consumed, smaller houses being utilized, fewer trips to the school and after-school activities, among many other reductions. And fewer children will mean that more parental and school effort and cost can be directed to achieving higher educational results and engaging in more high-enrichment activities such as travel, or attending cultural events such as musicals, music festivals, the opera, and the theater. This will be critical in future as robots and computer software will reduce the number of unskilled and semi-skilled blue and white collar jobs and place a premium on high educational attainment!

Reduce your driving! When and where available, rideshare, use public transport, live near your work, telecommute, and if you have to drive, buy a fuel-efficient gasoline or diesel-powered car or buy an electric car. We must stop driving our children to school. In an interview by Patrick Barkham, Mayer Hillman stated that, "in 1971, 80% of British seven- and eight-year-old children went to school on their own; today it's virtually unthinkable that a seven-year-old would walk to school without an adult." So we have removed children from a perceived danger and instead exposed them to a very real danger of roads full of polluting cars on early morning and late afternoon school runs. Hillman estimated that in 1990, some 900 million adult hours were spent taking children back and forth to school, costing the UK economy £20 billion each year. In the USA, fear mongering of child kidnapping by the print and TV media has had a similar effect.

We should also practice eco-driving, if driving an internal combustion engine car. We should keep car tires inflated, we should remove unnecessary and unused roof-racks, we should minimize the use of cars' air-conditioning units and, if available, use the engine-turn-off function when at stop lights, as this will save 5–8% of your fuel consumption. We would also do well to avoid sharp, aggressive accelerations and heavy braking, and aggressive driving in general.

Reduce or cease flying or fly at different elevations to reduce contrail-related warming! Greta Thunberg demanded that her parents stopped flying—her mother had to give up her career as an opera-singer. Or at the very least, demand that commercial airlines fly at levels that reduce the production of contrails. Although flying represents 5% of human-induced global warming, Stettler and coauthors suggest that 50% of this is associated with the formation of contrails caused by the emission of carbon black, and 80% of all "warming" contrails come from 2.2% of all flights either in the late afternoon or night where atmospheric conditions cause the persistence of contrails. Stettler and coauthors suggest that airlines need to fly lower in summer and higher in winter; although this would increase fuel consumption by 0.014%, it would reduce contrail production and hence warming by 58%. And if carbon black emitted by inefficient engines is reduced by transitioning to more fuel-efficient engines, Luntz and Stettler suggest much of the contrail contribution to climate change could be eliminated.

Do not eat meat! Or at least become a flexitarian! It is that simple! The IPCC suggests that by converting to plant-based diets, most countries would meet their IPCC 2018 GHG goals. In fact, 30% of global GHG emissions are the result of food production and food delivery. According to Matt Davis, if we were to substitute just 30% of each beef burger consumed in the USA each year with mushroom pate, this would be equivalent (in GHG emissions) to removing 2.3 million cars from US roads and eliminating water consumption by 2.6 million Americans.

We must also stop overeating; Rizvi and coauthors suggest that if the world followed USDA dietary guidelines, we would run out of land. They argue that if we are to follow dietary guidelines, these dietary/nutritional guidelines must be sustainable. In other words, they must consider plant-based diets.

And it is not just the production of GHGs that is worrying; according to Emery and Molidor, 80% of all US agricultural land is used to produce meat and dairy products, and 50% of all water use in the USA is consumed by the meat industry! Furthermore, according to them, if 500-seat catering events over two days were to offer earth-friendly, plant-based menus, each event would save five acres of land from animal agriculture, 10 tons of GHG emissions, and 96,000 gals of water.

But we should also stop rainforests from being cut to quickly generate organic soybeans and provide pasture for dairy and meat production in the tropics. We must, as citizens of this planet, make both more individuals accountable for their consumer tastes and food preferences and more businesses accountable for their operational activities in generating goods and services, as both have significant and damaging environmental consequences.

If vegetarianism or veganism is not for you, consider *eating more holistically*; in other words, eat more locally produced meat or farmed/ranched meat from animals natural to an ecosystem. For instance, replace beef in your diet with bison (*Bison bison*) meat. Prior to the emigration of European Americans into the Midwest of the USA, bison numbered in their tens of millions, but by the late 1800s, bison were nearly extinct as the then US government sought to undermine Native American livelihoods. Today, half a million bison roam private lands, ranches, and Native American lands. They were the keystone species of the American prairies: they fed on prairie grasses, which, without bison grazing pressures, grow quickly and outcompete other native plants. According to Klapp and coauthors, bison-grazed prairies are richer in biodiversity than cattle-grazed prairies because selective grazing of grasses by bison creates a rich mosaic of micro-habitats within prairies. For instance, bison prefer grazing new grasses that sprout up within recently burnt prairies while ignoring other areas. Bison also reduce woody species penetration of prairie due to their need to rub and itch themselves on woody shrubs and trees. Also, 10,000 years of bison manure was critically important to the production of prairie soils. Moran found that both bison grazing pressures and bison manure increase insect abundance and diversity in prairies. McNew and coauthors suggest bison were critical to the survival of the greater prairie chicken because greater prairie chickens utilize different micro-prairie habitats for displaying, feeding, and nesting. In other words, bison management for meat production is holistic and critical to the future conservation and richness of prairies.

Or if bison is not for you, reduce or eliminate your consumption of beef, sheep products, and dairy and become a flexitarian—these three food entities are the biggest producers of GHGs among all food products humans produce and the most resource-intensive, consuming huge amounts of land and water. In contrast, the production of rice, roots, tubers, and even corn is least polluting of all food products, whereas the production of beef emits 100 times more GHGs than the production of legumes. The farming of chicken, pork, and fish emits quite low amounts of GHGs; however, the consumption of marine fish and especially shark is completely unsustainable! A stark illustration of unsustainable fish consumption is that since the 1960s, the tuna industry has moved from catching and tinning bluefin tuna to yellowfin and albacore tuna, as their unsustainable fishing practices have all but eliminated commercially viable bluefin fishing. Today, bonefish restaurants are common in the USA, but 20–30 years ago, no respectable fish-eater would have considered eating this by-catch.

Farmed fish is a viable and sustainable alternative to industrial ocean fishing, but this has its issues. According to Fry and coauthors, half of all the world's consumption of fish is derived from farm-fishing. Farmed fish are increasingly fed, industrially produced terrestrial food. Therefore, any discussion of sustainability must address the use of pesticides, herbicides, and fertilizers used to generate the crop-based ingredients fed to fish in aquaculture.

We need to invest in green energy at the household level and, of course, among energy producers. This means investing in solar heating and electrical production and wind power.

We need to learn from Australia and conserve more water: this means installing low-flushing toilets and low-water-use washing machines and dishwashers, storing rainwater for use in gardens, infiltrating rainwater for drinking water, and using brown-water to flush toilets. Conserving water also reduces local, regional, and global energy consumption because providing clean freshwater to people is a major energy sink.

Low-impact actions that significantly reduce individual carbon emissions.

We must also eat more unrefined, unprocessed, organic, and locally and home-grown whole foods. Where available, we should pick our own food from the fields—this is a pretty common practice in Europe. We should shop more often and buy less food each time; this results in less food waste. We should also provide our own reusable shopping bags. We should follow the example set by the French National Government and legislate that restaurants should sell all uneaten meals either cheaply, give the food to homeless shelters, or just give the uneaten meals away after restaurant hours. We must also support organic, fair-trade farmers and NGOs that support them.

Grow as much of your own food where possible, and buy fewer stuff! We can also revert to old technology such as fruit walls that, between the sixteenth and twentieth centuries, allowed urban gardeners, nunneries, and monasteries to grow Mediterranean fruits and vegetables as far north as the Netherlands and the UK. Walls, be they brick or masonry, store solar energy during the day and release it slowly during the night, reducing the likelihood of late spring and early fall frosts. These sadly were replaced in the late nineteenth and twentieth centuries with greenhouses, but modern greenhouses require heating and so, in fact, consume energy. According to González and coauthors, 1 kg of bone-free beef requires 37–82 MJ/kg of energy consumption to produce and yields 20–40 kgCO_2eq./kg food; in contrast, tomato production in heated greenhouses requires 35–130 MJ/kg of energy consumption and yields 0.75–9.4 kgCO_2eq./kg food. This is quite disheartening and suggests we, as individuals, farmers, and society, need to rethink fruit and vegetable production as energy inputs in heated greenhouses are unsustainable.

Boycott goods, businesses, or governments that advocate unsustainable consumption or production We as consumers must start to flex our consider consumer power—we must, when and where possible, boycott products, companies, investments, even localities and nations, even if these entities are countries, if these *entities* are significantly contributing to GHG emissions, precipitating extinctions, and/or unsustainably exploiting natural resources. Ryan Leitner writing in the Global Nonviolent Action Database in 2014 illustrates how a successful boycott can work. In 1969, British college students formed a "Boycott Barclay" coalition to force Barclays Bank to divest itself of all investments in Apartheid South Africa. Beginning with a campaign to stop students from opening Barclays Bank accounts and also to opening fake accounts in 1970, to organizing branch-bank sit-ins in 1971, to disrupting Barclays AGMs in 1972, and beginning in 1973, pressuring organizations to close their bank accounts with Barclays. This latter effort started to pay dividends in 1980 when both Camden Council and the National Methodist Conference closed their Barclays accounts. In 1982, the Catholic Institute for International Relations ceased its relationship with Barclays. In 1984, students campaigned for the UN to support sanctions against both South Africa and those companies and countries that support or invest in South Africa. By 1985, Barclays' portion of the UK student market had dropped from 27% to 13%. In 1980–1985, it lost 6 billion pounds in revenue-earning accounts. In 1986, two more Oxford Colleges canceled their Barclays accounts, as did Oxfam and the National Association for Mental Health. The following year, Barclays retracted all investments from South Africa.

On the other hand, we must also embrace, use, and purchase "green products." These can include sustainable and renewable energy sources, organic food items, and recycled products, and use public transport. And when and where possible, buy local products.

We must reduce our consumption of clothes and new furniture In the Global North, people are buying 60% more clothes in 2014 than they did in 2000, and this trajectory is not changing. And according to McFall-Johnsen, most of these clothes are manufactured in the Global South or China and this relocation of clothing manufacturing has resulted in a massive increase in GHG emission by the fashion industry to the point that the industry contributes nearly 10% of all GHG emissions. The fashion industry also consumes huge volumes of water and pollutes streams, rivers, lakes, and the oceans with thousands of microfibers.

We must learn to sew again, to replace buttons, and to darn socks. We also need to buy local to reduce the emission of GHGs in shipping and, where possible, buy quality that lasts over quantity that falls apart easily and quickly.

We must also wash our clothes less! Blue denim jeans dyed with indigo, a natural antibacterial agent, do not need to be washed regularly. Rachel McQueen found that a pair of jeans worn for 15 straight months without washing exhibited the same bacterial load and odor as a similar pair that had been worn straight for 15 days without washing.

We should buy more vintage furniture, especially sofas, love-seats, and chairs that provide support through woven strapping and springs, and avoid buying new overstuffed furniture full of polyurethane foam that are significant contributors to household VOCs (volatile organic compounds) and that contribute to a range of health problems, from allergies to cancer.

Plant trees and shrubs in your own front and back gardens and along roads and stop maintaining a lawn Lawns are biological deserts and mowing contributes significant GHG emissions. According the British Ecological Society, one immediate way to increase invertebrate and plant diversity in the built environment is to mow lawns less frequently: they found that even a modest reduction in lawn mowing increased the presence of pollinators, increased plant diversity, decreased the presence of ragweed (a disturbance specialist that thrives in intense mowing environments and major source of allergens in the built environment), and reduced GHG emissions. According to Watson and coauthors, the city of Trois-Rivieres, Canada, found that by reducing mowing frequency by a third led to a 36% savings in public maintenance costs. In the USA, $20 billion dollars are spent each year on allergy relief, and it is estimated that 15–25% of the US and European populations suffer ragweed-allergies. So any reduction in ragweed-allergens would be fiscally prudent in several ways. We, as house owners and lawn-care specialists, have to just accept longer grass, or just go further and eliminate lawns altogether, as the dry-garden movement in the American South-West is trying to accomplish.

We should also design gardens that are bee-friendly and drought-resistant; this means more gravel and stones and native shrubs, plants, and trees; compost unused food, grass cuttings, and other yard waste; mow your grass and yard less frequently and avoid mowing things like dandelions, daisies, and buttercups by creating paths of regularly mowed lawn surrounded by more rough seasonally cut grass that offer places for native grasses and plants to flourish.

Grass-verge maintenance along and between roads needs to be severely curtailed! The state of Georgia in the USA just removed much of the trees within the median of I-16 from Macon to Savannah, replacing the pines with grass. The long-term increase in maintenance costs and increased GHGs will be substantial and could have been avoided. Recent work by O'Sullivan and coauthors suggest road-verges offer significant ecosystem services and greatly enhance urban biodiversity. They recommend reducing mowing frequencies, planting diverse and forb-rich species mixes, planting native shrubs and trees, retaining mature trees in the verges, and increasing tree species diversity.

Maintaining and increasing the number of urban trees has a significant impact on urban heat islands. In Montreal, Wang and Akbari found that at midday, at 20 m above the ground surface, trees reduced temperatures by 4 °C and at 60 m by 2 °C.

We must live and work in higher density urban environments Why? We must conserve what remaining space is left outside our sphere of impact for other animals and plants. We must begin to generate food in three-dimensional space! We must also support and embrace urban architecture that reduces GHG production, generates electricity, and creates more shade.

Traditional Mediterranean buildings and roofs are painted white to reflect solar radiation and reduce heat gain, are positioned closer together to generate more shade in lanes, alley-ways, and roads that bisect the urban built-environment, and frequently rely on open rooms facing an interior unroofed courtyard with a small pool to increase ventilation (◘ Fig. 9.1); Iran and Qatar utilize wind-catchers and pools to ventilate and cool rooms.

We must stop taxing domestic solar panels across the USA and support their installation and encourage the use of domestic-house scaled windmills. If William Kamkwamba, a young teenager in famine-ravaged Malawi, can build a cheap electrical-generating wind turbine out of a bike and junk, so can we!

We must demand our local, regional, or national governments establish and enforce "green housing codes" that require housing to be more fuel-efficient and that all accommodations must also be passive solar and must have wind- or solar-generated electrical generators. In addition to solar and wind electrical generators, we must insulate our houses more holistically using sod roofs, straw-bale walls, and heat sink designs. We must also use more recycled steel and old timbers—and do as the opal mining town of Coober Pedy has, bury itself underground. This reduced cooling costs in summer and warming costs in winter, and reduces the risk of storm and fire damage. We must redesign houses to be square or round to reduce heating and cooling costs. We must reduce the size of our homes, particularly in the USA.

We must encourage German-style sustainable cooperatives. In October 2006, the residents of Großbardor in Rhön-Grabfeld county in Bavaria, Germany, established Agrokraft—a cooperative that plans, initiates, implements, and optimizes renewable energy projects and sustainable agricultural projects with the goal of developing rural areas and indirectly reducing brain-drain and population flight to the cities. The Großbardor projects preceded the German federal government's efforts at *Energiewende*. By the way, *Energiewende* is a planned federally -assisted transition by Germany to a low-carbon, environmentally sound, reliable, and affordable energy supply. As part of this effort, the federal government of Germany announced in a 40 billion euro agreement on January 16, 2020, that all coal-fired power plants are to be shut by 2038.

Agrokraft and Großbardor's first project was a 7.6 million euro solar panel power plant whose peak power output is 1.91 MWp. This was submitted by 100 initial shareholders and supported through numerous low-cost federal loans. Their second project created a grandstand with a roof covered in solar panels for their football team—these panels produce 125 kWp. Recent projects include a district

9

Fig. 9.1 The central courtyard of Gayer-Anderson Museum, Cairo, Egypt. By Peter Astbury – Own work, CC by 3.0.

heating network for Großbardor and biogas plant that generates 625 kWp and 680 kW and the building of four windmills. Today, Großbardor produces 475% of its electrical needs and sells the excess, and generates 90% of its heating needs.

9.4 We Must Support NGOs that Promote Sustainable Food Production and Conserve Natural Resources

There are a myriad of NGOs that purport to foster environmental and natural resource sustainability, in other words, are green! Please do your homework before committing funds to them! Not all are as green as they advertise! We recommend looking at their bottom-line, in other words, how much of their donated funds go to their advertised causes. High overhead is always concerning.

Environmental Non-Governmental Organizations (NGOs)
Although we do not have the space to recommend all local, regional, and national organizations, some international NGOs are outstanding and worth mentioning:
- BirdLife International
- American Bird Conservancy
- Royal Society for the Protection of Birds (RSPB)
- World Wildlife Fund
- CERES
- Conservation International
- Greenpeace
- Natural Resources Defense Council
- The Nature Conservancy
- Oxfam
- Slow Food International
- Environmental Defense Fund
- Center for Biological Diversity

We must support NGOs that collect, collate, store, analyze, and make available worldwide biodiversity data Among the most important NGOs are Cornell Lab of Ornithology and their database—eBird. Developed in 2002 to promote citizen science and the collection of environmental data, eBird failed! Their 2002–2005 slogan "Birding with a purpose" was replaced with "Birding in the 21st century." Caren Cooper argues that Cornell Lab for Ornithology and eBird realized that providing birders with better tools to collect basic environmental data was infinitely better and more successful than trying (and failing) to appeal to birders' ethical sense of environmental duty. eBird also encouraged birders to submit complete checklists from each location they birded. This generated more usable scientific data. eBird also understood that the 90-9-1 rule of online discussion forums play a significant role in citizen science data collection as witnessed by Wikipedia. For instance, 90% of users just browse for information, 9% edit information within the forums, and only 1% create new content. To overcome this, eBird online and their app are geared to show through maps, alerts, and target bird analytics where birders might find new birds already logged by other birders. According to Cooper, by 2013, eBird stored over 140 million observations, 2.5 million people had engaged with eBird, 150,000 (6%) had submitted data, and 25,000 birders (1%) had submitted 99% of the data. And some 90 peer-reviewed articles had used eBird data. As of January 1st, 2020, eBird stored 737 million observations. In 2019, 5.2 million visited eBird online, 500,000 eBirders submitted data (9.6%), over 100,000 people downloaded raw data for analysis (2%), and since 2002, over 300 peer-reviewed articles have used eBird data. Today, eBird is the world's largest biodiversity-related citizen science database.

And we must at times embrace radical solutions! For instance, in a preemptive move to stop the extinction of the Lord Howe Island's woodhen and currawong, all surviving birds were captured and housed until the largest rodent eradication program ever attempted was completed. This joint project venture was led by Lord Howe Island Board, Sydney's Taronga Zoo, and the state of New South Wales (Australia) Department of Planning Industry and the Environment. According to Siossian and Wyllie, all woodhens and currawongs have been subsequently released back into the wild, and as of January 2020, both were once more breeding.

9.5 What Can *Groups* Do to Mitigate Climate Change, the Sixth Mass Extinction, and Natural Resource Exhaustion?

We must as educators and the media address divisive racial attitudes toward climate change. Sadly, several large opinion polls in the USA illustrate, according to Salil Benegal, that individuals that exhibit high racial resentment, those that are white and old, are more likely to deny human-caused climate change. Amazingly, he also found that Republican and independent voters will accept human-caused climate change if Republican legislators vocally support climate change.

We must as scientists, planners, legislators, resource managers, architects, NGOs develop practical solutions to mitigate climate change, for instance, by increasing the capacity of carbon sinks; hence, we must support efforts to reforest and eliminate high GHG emission land-uses, increase ecosystem services, and stop extinctions.

We must increase fuel efficiencies among jet and internal combustion engines—thankfully the increasing cost of jet fuel, petrol, and diesel are forcing engine manufacturers to increase efficiencies. Among internal combustion engine manufacturers, developing hybrid drive systems utilize energy return systems (ERSs) that allow the cars to recover energy typically lost during braking by converting momentum into energy that is stored and then used to boost power rather than letting it go to waste. This makes the car more fuel-efficient. Formula One racing cars are at the cutting-edge of this technology. Formula E racing is also pushing the limits of electrical battery storage and electrical motors. Elsewhere, Norway has legislated to reduce internal combustion engine car purchases by removing a 25% sales tax (in 2001) from all new electric vehicle purchases. Today, greater than 60% of all new car sales are electric vehicles. This is despite Norway remaining a major oil producer.

We can work with groups, among many other efforts, to agitate and legislate to ban trophy hunting; ban the shooting and trapping of migratory birds on Mediterranean islands; stop badger and fox hunting; stop vulture poisoning across Africa; ban illegal pet trade; ban bushmeat consumption especially among animals that potentially harbor emergent infectious human diseases, that is, bats; support GreenPeace and ban whaling; and ban shark finning!

We can also encourage skyscrapers to turn off lights during bird migrations; promote communities to encourage and facilitate the nesting of bats, taking inspiration from San Antonio in Texas; support BirdLife International efforts to force long-line fishing communities to use weighted lures that significantly reduce the bi-catch of albatrosses; enforce illegal trade in animal parts that are pushing rhinos, elephants, tiaga antelope, helmeted hornbills, and pangolins to the edge of extinction; increase the number and security of tropical parks; reduce tropical and boreal deforestation; ban the use of the neonicotinoid family of pesticides as the France as done; ban one-use plastics; promote and implement the elimination of invasive predators from islands and continents taking inspiration from the work that New Zealand and Australia governments and many NGOs are currently engaged in; eliminate micro-beads in consumer and industrial products; ban the encroachment of mining into conserved wild land areas; and among many other things stop the encroachment of low-intensity agriculture and urban sprawl into wild areas.

In addition, we must follow the lead of Germany and begin shutting all coal-powered electric-generating stations. We need to emulate Portugal's green power production—currently Portugal produces more than 50% of its electricity from renewable sources, most through wind turbines. In 2016, the Pacific Islands Renewable Energy Investment Program was announced by the Cook Islands, Tongo, Republic of the Marshall Islands, Federated States of Micronesia, PNG, Nauru, and Samoa to achieve independence from diesel power by converting to solar, hydropower, and wind energy. Even before this announcement, by the end of 2015, the northern atolls of the Cook Islands were fully solar-powered.

We as scientists and environmentalists need to emphasize the importance of ecosystem services for the future survival of our species! We cannot continue to survive on this planet without the continued functioning of a myriad of ecosystem services. We must legislate to support them. According to the FAO of the UN, these include the following ecosystem services: providing nutritious food and clean water; regulating disease and climate; supporting the pollination of crops and soil formation; and providing recreational, cultural, and spiritual benefits. Together, these asserts are worth $125 trillion, yet these are not adequately accounted for in political, economic, or agricultural policies worldwide. As a result, their protection and management is minimal at best, and at worst ignored.

Academics must stop talking just to academics Scientists need to engage both the general public and laypersons and our legislators. For too long scientists have just talked to each other. As Wu so simply states, most scientists tend to "other" the general public and in doing so create an unnecessary hierarchy. In other words, othering makes non-scientists look different, distant, and second-class. By othering non-scientists, scientists make effective communication between the two groups difficult and, as it turns out, very detrimental to the environmental mitigation. This othering has been compounded by the politicization of science. Public trust in science, particularly in the USA, is at a low ebb. Anti-vaxxers and climate-deniers are by-products of this politicization and othering—incredibly, 97% of all scientists believe in climate change (IPCC), but as of late 2019, more than 20% of the US population do not, and we saw similar numbers in Australia five years ago. Recent polls suggest Australians are changing their minds. In April 2019, according to an Ipsos survey, 46% of Australians accept that climate change is human-caused, up from 32% in 2011. We suspect that unpresented six-year drought might have changed Australian's view of climate change before the bushfires of late 2019/early 2020.

9.6 Recycling, Energy Usage, and Alternatives

Although recycling continues to be an individual-level commitment to combating environmental degradation broadly and helping mitigate climate change specifically, multiple concerns have emerged recently, many of which are related to energy usage in recycling processes. These have led to considerations that governments and non-state actors alike must address if recycling is going to continue to be an environmentally helpful practice.

Plastics Petroleum serves as the source of the vast majority of plastics created, which, of course, means they are not biodegradable and thus accumulate in landfills and the natural environment. Indeed, the same characteristics that make plastic an attractive material, durability, for instance, makes it a veritable nightmare for the environment. There are only three fates for new plastics once they are created (often referred to as "primary" plastics). They can be disposed of in landfills or the natural environment, they can be recycled (the result of which is referred to as "secondary" plastic), or they can be destroyed. The only way to destroy most plastics is through some type of combustible process (e.g., incineration), and of course plastic recycling is a relatively environment-friendly alternative to disposal in landfills or incineration. However, according to a 2017 study, globally only 9% of all plastics ever produced (dating back to the 1950s) have been recycled, while 59% ended up in landfills (the remaining 32% are either still in use or were incinerated). Given that plastic recycling has not always been an option, the good news is that the global recycling rate has increased to 18% today. However, in the USA, only 9% of plastics are recycled, while 75% are discarded; these rates have remained steady in recent years. Plastic recycling rates in the EU and China, by contrast, are approximately 40% and 30%, respectively. Although plastics are not biodegradable, sunlight does weaken plastic material over time, leading to smaller and smaller fragments of plastics finding their way into the environment. Research on microplastics (defined as smaller than 5 mm in diameter) is fairly nascent, but studies have shown their presence in freshwater and oceans has negative impacts on fish and other aquatic life.

The largest single use of plastics is packaging, with approximately 42% of all non-fiber plastics ever made. As we have discussed previously regarding environmental issues, things are expected to get worse, not better, as more and more Global South countries achieve higher levels of development. Specifically, as consumption levels increase in Global South countries, the amount of plastic created (and thus requiring disposal) is only expected to increase. Some alternatives to petroleum-based plastics do exist, and such bioplastics may help alleviate some of the disposal issues described earlier. However, given that a 100% transition to bioplastics would require nearly 5% of all arable land on earth, other options or solutions must be considered if we are to at least slow the growth of plastic waste.

Metals, rare earth elements, etc. The demand for rare earth elements needed for large-scale decarbonization of energy production (e.g., dysprosium, europium, and terbium) will only increase with growth of cities in Global South, and recycling and reuse of rare earth metals must improve (in both quantity and quality). In addition, studies have identified several critical and near-critical metals (e.g., gallium and graphite) for a low- or zero-carbon energy transition, while market incentives for recycling and reuse will not be sufficient to meet future demand. Thus, international policy alternatives including setting prices high for some of these materials might be needed to meet the needs of a decarbonized global energy system.

Immense opportunities exist for low- or even zero-carbon energy uptake for metals recycling, since recycling metals requires less electricity than does metal production from raw materials. So, the potential for electricity usage via biomass or solar energy in metals recycling is great.

Our Western-Disposable Society: The Diaper, the One-Use Plastic Straw, the Plastic Coffee Cup, the Plastic Water-Bottle, the Plastic Shopping Bag

Diapers. Nothing else defines our current Western throwaway society than the disposable diaper—easy to use, easy to dispose of but catastrophic to the environment. The 27.4 billion disposable diapers thrown into US landfills each year represent 3.4 million tons of waste or 2–4% of solid landfill waste. From birth to a potty-trained kid, a period of 2.5 years, a single kid's diapers will consume 20 trees and 7 barrels of petroleum.

In a report released by the U.K. Environmental Agency in 2006, in 2.5 years, a single kid and their disposable diapers will release the same amount of GHGs and consume nonrenewable resources as driving a car at least 1300 miles. In other words, each year in the UK, disposable diaper-wearing kids emit GHGs and delete nonrenewable resources equivalent to the consumption and emission of 98,600 cars, each driving 12,000 miles. Decomposing disposable diapers also releases CFCs that contribute to ozone depletion.

In contrast, a home-washed cotton cloth diaper has half the ecological footprint of disposable diapers, although washing cotton diapers at home typically uses 230 liters of water every 3 days (Lehrburger 1991). Yet even from a baby's health perspective, disposable diapers make no sense; diaper rash nearly unknown in the 1950s afflicts nearly all babies wearing disposable diapers. However, if parents participate in cloth diaper delivery services, the carbon footprint of prefolds (i.e., the CO_2 produced in the delivery of fresh cotton diapers and pick-up of used cotton diapers) is slightly higher that production, distribution, and disposal of one-use disposable diapers.

Nevertheless, a *Washington Post* article from 2015 found that most cotton, except that produced in the southeastern USA, requires extensive irrigation. In fact, 30 cloth diapers use 1221 m^3 of water versus just 141 m^3 of water in the manufacture of 4000 disposables (Salcito 2015). In addition, 54% of India's annual pesticide use is sprayed on its cotton crops, even though these crops cover just 5% of India's cropland.

So, given conflicting statistics, are cotton diapers "green"? How do ethical consumers balance emission of GHGs and consumption of nonrenewables in the production, distribution, and disposal of disposables with water consumption and pesticide use in the production of cotton and rewashing of cotton diapers? Do ethical consumers in the end fall back on the fact that cotton diapers, for all their problems, are better for babies, causing far few incidents of diaper rash? In causing less diaper rash, do cotton diapers then reduce the nonrenewable resource consumption and emission of GHGs inherent in the production, distribution, and use of diaper rash products that are manufactured using petro-chemicals?

One-use plastic straw. The use of plastic straws is ubiquitous among the world's restaurants; sadly, as a result, some 8 billion plastic straws litter beaches worldwide. In fact, some 500 million one-use straws are produced in the USA alone. Yet we can easily do without a drinking straw.

Recent efforts by environmentalists and consumer and conservation groups have been successful in lobbying large consumers of plastic straws to limit their use, propose to stop using them in the near future, or stop using them now. However, is the elimination of plastic straws a *straw dog*? Have these consumers—for example, McDonald's and Starbucks—given up something that is easy to eliminate and not costly to replace to get good Public Relations, while still using other one-use items such as one-use plastic cups that are expensive to eliminate?

One-use water plastic bottle. Thankfully, ethically conscious water consumers have been using reusable Neogene bottles for decades; however, 45 billion plastic water bottles are still made in the USA each year. Many US citizens are still convinced that bottled water is cleaner than tap water; however, the EPA does not regulate water that is sold in plastic bottles. Yet most water in plastic water-bottles is tap-water. In fact, thousands of plastic micro-particles occur in every plastic bottle of water, and a third of all plastic water-bottles test positive for harmful bacteria. For every 1 L plastic water-bottle produced, 62 g of GHG are released and 90 g of petroleum and 4 L of water are consumed. Some 17 million barrels of oil are used in the USA to manufacture plastic water-bottles. Today, discarded plastic water-bottles occupy in excess of 120 million m^3 of landfill, yet fewer than a third of all plastic water-bottles are recycled in the USA.

9

Recycling plastics

Recycling is a business. Egyptian Copts have cornered the recycling market in Egypt, but their recent success has fueled inter-ethnic and religious rivalries and sparked anti-Coptic riots. But it is a business that suffers from boom-and-bust cycles dependent on the cost of petroleum products. When oil prices are high, recycling plastics is profitable; however, when fuel prices are low, recycling is no longer profitable. In the USA where petroleum prices are much lower than elsewhere in the world, plastic recycling is barely profitable locally, so nearly all plastics are shipped to China.

As a result, China dominates recycling, in particular, buying used plastic products from across the globe. However, in early 2018, the Chinese, after repeated warnings issued to the USA, refused to buy anymore plastic from the USA because of persistent food-contamination of plastic garbage. They will only accept narrow opening plastic bottles. Unlike the EU where citizens can and are fined for mixing their recycling and not separating food waste from plastic or washing out their recycling plastics, few municipalities in the USA exercise such draconian policies. Most US municipalities simply cannot afford to monitor plastic recycling because they are already at their break-even profitability point. Consequently, in the USA, plastics contaminated with food waste is the norm.

Even car recycling is now a global enterprise. The LYNXS automotive shredder in Newport, Wales, digests up to 450 cars per hour. The shredder separates the plastics, rubber, carpet, and glass for local recycling, while 2.5 million tons of metal scraps, mostly copper, iron, and aluminum are shipped to East and SE Asia annually.

Recycling Ships

Even ships are recycled, but in the ship breaking yards of Bangladesh, India, China, and Pakistan that employ nearly 230,000 workers, environmental pollution is significant and uncontrolled. The three largest centers are Chittagong in Bangladesh, Alang in India, and Gadani in Pakistan. Chittagong yard generates 20% of Bangladesh's steel needs. Incidentally, the Chittagong yard started completely by accident when in 1960 a cyclone beached the Greek ship M D Alpine. In 1965 the ship was scrapped in situ.

First, all fuel and liquids are drained and sold. Second, all wire, furniture, and usable machinery are stripped and sold locally or to the global trade markets. Insulated wire is stripped and the insulation burnt off. The steel hull is then cut into blocks and 90% of it recycled as re-rolled scrap. Painted steel is re-rolled liberating dioxins. Most hazardous materials are stock-piled on the adjacent beaches and left. Soil and beach contamination at Chittagong with iron, chromium, nickel, zinc, arsenic, lead, copper, cadmium and mercury exceeds all known safe levels. Chromium, lead, cadmium, and mercury are particularly toxic and known to be carcinogenic, teratogenic, and neurotoxic. At Chittagong, as of 2010, some 79,000 tons of asbestos, 240,000 tons of PCBs, and 210,000 tons of polyurethane foam lie in permanent storage (see Sarraf et al. 2010 for more details). Leaching of heavy metals, oils, and PCBs into the Bay of Bengal is unconstrained and hazardous. Future sea-level rise due to climate change will also increase the volume of polluted sand exposed to tidal processes. Yet the coastal and marine fisheries in the Bay of Bengal are contaminated by these bio- and geo-accumulating heavy metals; however, at present, only arsenic represents a serious human health risk as a carcinogen (see Saha et al. 2016 for more details).

9.7 What Can *Legislators* Do to Mitigate Climate Change, the Sixth Mass Extinction, and Natural Resource Exhaustion?

First, we as *individuals* and *groups* have to convince *legislators* that they must act, and act immediately. Many governments are paying heed to environmental-society needs. For instance, the German Federal Government announced in early 2020 it would cease to use coal in electrical generation by 2038! This is a start, but giving an 18-year leeway is just too long. In contrast, the Chinese government announced ceasing the use of single-use plastic bags by the end of 2020 in large cities, and all other environments by 2022. We are comparing apples-and-oranges here, but immediate action is needed.

For instance, Sweden's commercial airlines are seeing a sharp drop in passengers—there has been a drop of 2 million domestic passengers between 2018 and 2019 or 9% of their internal passenger traffic according to Swedavia. Analysts suggest a new Swedish aviation tax, poor economy, weak Swedish crown, the climate debate, and flight-shaming. As a result, Swedes are traveling more by train than ever before. In contrast, the EU saw an additional 100 million passengers flying within the EU, and 10% of these numbers were accounted for within the UK.

9.8 Looking Back, to Look Forward

The simultaneous occurrence of climatic change, epidemic plagues, and carrying capacity exceedance leading to the collapse of a society or civilization has not to our knowledge occurred before. Nevertheless, Mayan and Anasazi civilizations and Norse Greenland islanders were felled by climate change coupled with carrying capacity exceedance, although diseases could have also played a knockout role. In addition, the Norse Greenlanders and Henderson/Pitcairn Islanders also lost their sustaining trade partners. Easter Islands collapse, still highly debated today, appears to be a combination of isolation, carrying capacity exceedance, and political misadventure. But in all cases, environmental destruction of natural resources, absolutely necessary for life, perpetrated by these societies, continued right up to their collapse: Easter Island and Iceland—the deforestation of their lands; in Greenland—the burning of grass sod for fuel.

However, climate change coupled with a global pandemic, and arguably carrying capacity exceedance, did occur at the time of the primary wave of the Medieval Black Death in 1347–1351 in Europe. This global pandemic also killed millions from the Middle East back in an arc to central Asia and China. The primary wave was preceded by an enormous famine in 1315, suggesting that agricultural technologies and population-carrying capacities were at a breaking point. At this point in history, the prior Medieval Warm Period (950–1250) was ending due to the failure of the deep Atlantic Conveyor and North Atlantic Drift as freshwater flowing off the melting Greenland ice cap shut down the sinking of hyper-saline waters off the coast of Greenland. The Black Death pandemic appears to have fueled this cooling as millions of acres of farmland underwent secondary succession back to forest trapping and sequestering millions of tons of CO_2. The primary wave killed 35–55% of the European population, and by the time the Medieval Black Death disappeared in 1801, as many as 75 million had died.

Yet this brutal, Malthusian combination of climate change, global pandemic, and initial population-carrying capacity exceedance radically changed the social, economic, and political conditions across Europe. Millions perished, but millions survived—others survived the MBD because of collective community decisions to quarantine strangers, something that seems to have happened in an organic way in more isolated rural villages in the Pyrenees.

The millions in Europe who survived this period from 1347 to 1801 witnessed and participated in some of the greatest changes in human history—the Renaissance, the Enlightenment, the British and French Revolutions, both the Agricultural and Industrial Revolutions that initiated the transition from animal power to stream power, the move from wood to coal energy resources, and the rediscovery of concrete. Europe went from the Dark Ages to the Age of Colonialism, the precursor to our system of multinational corporations and globalization. But it is important to note that the immediate period after the primary wave of the Black Death was racked by widespread violence, lawlessness, widespread distrust of church hierarchies, and political upheaval. But even here, Feudalism died, while towns and cities and their gilds, the precursors to modern corporations, grew, and land was swapped for money as a sign of wealth. Many rich families disappeared, while the poor suffered the highest mortalities, and some survivors from the rural and urban underclass rose to wealth and prominence. In fact, the British class system, rather than being immanent, is highly fluid, suggesting the rise and fall of families across generations is the norm—wealth, power, and privilege, in fact, rarely seem to survive more than three generations.

9.9 Overcoming Looming Global Challenges

The juxtaposition of the major challenges in the previously discussed triumvirate represents a major challenge to the current Western standard of living. Previous works have addressed individual components of this triumvirate and the issue that may follow: myriad works on the end of the oil age, on the issues that could potentially occur due to climate change, and on the food and clean water issues confronting the global population, including a multitude of additional challenges in the developing world. But these issues are not new—in 1972, the Club of Rome, a global think tank, published *The Limits to Growth* (Meadows et al. 1972). Their modeling identified that rapid population growth, coupled with depleting energy and other resource supplies, and increasing pollution, would lead to a global population collapse by the middle of the twenty-first century. Although widely criticized at the time, subsequent models and analysis seem to confirm that changes in food production, energy and industrial production, and pollution (although not explicitly CO_2 and global warming) since 1972 do confirm that societal collapse will occur in the twenty-first century if our global society maintains a "business as usual" model of resource extraction, use, and resultant pollution! *The Limits to Growth* offered several remedies to avoid collapse.

These suggestions were for governments across the world to put immediate limits on population growth and pollution generation and cease all economic growth! Of course, these draconian-level responses have not occurred at the country-level. Bipartisan cooperation is difficult enough in the US House of Representatives and US Senate, let alone across the globe. The 1992 Kyoto Protocol on climate change was ratified by most of the nations in the world except those with most to lose—the

USA and China. The initial recommendations laid out in *The Limits to Growth* were echoed by Graham Turner in 2008 when he reanalyzed the models with 30 years more data (Turner 2008). The trends identified in 1972 were still present and were repeated in the most up-to-date models—global socioeconomic collapse by 2050. Simply put, the world economic system, as we know, is on an "unsustainable trajectory" that is a one-way street unless there is a massive global reduction in consumption. However, growth in consumption and economic production is what drives the world economic system—stagnation is unacceptable. Therefore, the solutions to the myriad problems affecting the global biosphere are the antithesis of the economic foundations and goals of everyday life for most of the people in the world, especially the developed world.

Although the picture admittedly appears bleak at first glance, we posit that such challenges are not unprecedented and that deliberate techniques have worked to prepare for and respond to such civilizational challenges. A good example is Europe's rather belated response to the world conflicts that erupted twice on that continent. Following WWI, Europe and the rest of the world tried to develop cooperation across divided Europe to forestall any future conflict. However, the League of Nations had no real power, and the Allies were bent on extorting Germany for war reparations, ultimately bankrupting Germany and leading to the rise of Nazism and Adolf Hitler (Shirer 1990). After their miserable post-WWI failure, Germany, Belgium, Italy, and France were in no mood to replicate these events following WWII, and set Europe on course to the European Economic Community and the European Union. In particular, Jean Monnet, architect of the ECSC and EEC, understood that fundamental and transparent social, economic, and political cooperation across nations stood at the foundation of any attempt to survive or avoid any political and/or economic disaster (Monnet 1978). Monnet persuaded six governments—Italy, West Germany, France, Belgium, Luxembourg, and the Netherlands—that they must adopt a radical new approach to European politics: cooperation, not confrontation. So far, Monnet's creation has, at least within Europe, stopped another continent-wide conflagration. In a sense, the EU saved the western European civilization. Today, the EU is the largest economy on earth (Rifkin 2013). It also has the strictest environmental regulations and sponsors the broadest conservation measures across the planet. It legislates to maintain public transport and funds the development of high-speed rail connections among its many attempts to reduce community member CO_2 emissions. Although there is a siren call to peoples across the world to "think global, act local" to address issues of rampant, unsustainable, unethical globalization and resource depletion, at least the EU might offer EU member-states and citizens an organization possibly capable of coordinated planning and preparation for Kunstler's "Long Emergency" at the supra-national level.

In the simplest sense, the first step in the long road ahead will be to save and conserve as many assets as possible. By "assets," we refer to the items required in daily life, and therefore, not necessarily the economic definition of asset, which are tangible or intangible items that possess the potential to increase in value. Certainly, items related to our standard of living have the potential to increase in value, but their role as an asset as used here is fundamentally related to their usefulness for a particular purpose.

Fig. 9.2 Prejmer, Romania.Welford @ 2003

A recent, historically verifiable illustration of these ideas is now known as Romania. Beginning in the twelfth century, the Saxons colonized Transylvania. Following attacks by Tartars and Turks, more than 300 Saxon churches were fortified against the invaders. Today, good examples of these fortified structures remain, and seven are now listed as UNESCO World Heritage Sites (Biertan, Viscri, Saschiz, Prejmer, Valea Viilor, Calnic and Darjiu). At Prejmer, the town's council required that each household store a portion of their harvest in one of 272 provisional chambers within the fortified church wall (Fig. 9.2).

Each household was designated a chamber to shelter in and store their food. During times of frequent crisis (in fact, this very church was besieged ~50 times!), the farmers retreated to the church and lived off their stores.

Therefore, as illustrated by the fortified churches of Transylvania, the second step in adapting to the challenges ahead involves the proper utilization of accumulated assets to respond to crises as they arise. In the case of the Saxons of Transylvania, it was the use of surplus food stocks to survive a scarcity in precious food stocks. This may very well be necessary in our future also! However, other accumulated assets may include items such as gasoline and fossil fuels, LNG, energy-storage devices (batteries), water, beverages (including alcoholic beverages), security-related items, and many other commodities of modern life. As with the term assets, "commodities" has a specific definition within the field of economics. Here, we refer to commodities as items that are designed to be used for a specific-purpose and have a finite lifespan. Henceforth, we distinguish between items that may be considered assets as only those items that may also be considered commodities, and therefore central to survival or maintaining the standard of living that is expected in the Western world.

The first two steps in the adaptation are therefore preparation (and the many aspects which coincide with this phase) and response (which consists of several additional phases). The final step in the adaptation process is evolution. An evolution occurs when a civilization adapts to the changes confronting that civilization in a permanent manner, and in which a reversion to "the way it was before" is neither anticipated nor expected. One excellent example of evolution, which is discussed in detail later in this chapter, is the evolution in energy supplies during the mid-nineteenth century involving a transition from sperm whale oil energy supplies to coal and gas-based energy supplies. Although the populace had been used to using whale-oil lamps and whale-oil heaters for several decades, the time came when the whales (as a natural organism and as a commodity) were almost hunted to extinction. A transition to a new energy source was necessitated because the *supply* of the old energy source was dwindling and nearly wiped out, in addition to the environmental degradation that was occurring through the hunting of the sperm whale. Another example of such evolution occurred just prior to WWI. The British Navy had converted to oil-powered ships, whereas the Germans entered WWI with a fleet of coal-fired vessels (Kunstler 2007). Consequently, the British fleet was more powerful and faster and had greater range, and resupplying the fleet was cheaper, less dirty, quicker, and required fewer people. The result was a foregone conclusion—the British surface fleet reigned supreme across the oceans during WWI. It was only challenged by German U-boats which, interestingly enough, ran on oil. In many ways, this issue parallels our current conundrum—our supply of fossil fuels is finite and is running out, while at the same time the environmental damage caused by the use of these fuels is magnifying. Unfortunately, the energy transition that awaits us is not a matter of convenience as was the transition from whale oil to fossil fuels. Today, our entire society depends on these fuels for its unsustainable standard of living.

Robin Mills, in his work "The Myth of the End of Oil," has defined a group of similar-thinking individuals as the Neo-Luddites—those advocating a lower-tech, slower-growth type of lifestyle that would be better prepared to handle the coming triple-shock of energy resource depletion/substitution, climate change, and carrying-capacity exceedance via "back-to-the-earth" lifestyles and common-sense sustainability measures. Is this really the direction that we must pursue in order for modern societies to survive and thrive in the coming decades? Or perhaps, we will have to adopt a more carefree and less consumer-oriented society such as that advocated by the "hippies" of the 1960s and 1970s—although, as has been noted by many authors, these seemingly Earth-Worshipping souls were driving around the country in gas-guzzling minivans, spray-painted with chemical-industry-created paint filled with VOCs and oil-based colors, and so forth. Surely, there must be a happy medium between a completely Luddite form of existence and the entirely unsustainable manner of modern Western life.

9.10 Who Is Going to Survive the Crises of Climate Change, Oil Shortages, and Environmental Services Collapse If We Do *NOT* Adapt or Mitigate?

In facing a crisis—we cannot predict who will survive, but rather individual survival, community and collective survival are contingent on many things such as political adaptability; personality traits such as height, weight; economic position; access to immediate disposable income; educational level; and connectivity with and outside a community. So, there is order within chaos! Stephen Jay Gould when discusses early life and anatomical diversity had this to say:

> » The new view, on the other hand, is rooted in contingency. With so many Burgess (Shale) possibilities of apparently equivalent anatomical promise – over twenty arthropod designs later decimated to four survivors, perhaps fifteen or more unique anatomies available for recruitment as major branches, or phyla, of life's tree - our modern pattern of anatomical disparity is thrown into the lap of contingency.
> Stephen Jay Gould – Wonderful Life

Local, regional, national, and global economic bubbles have plagued humanity for millennia, for instance, the Tulip mania of the 1630s in Holland, the narrow gauge railroad movement of the 1870s in the United States, towns such as Bodie, California, that prospered during the California Gold Rush of the late 1840s and early 1850s then became ghost towns, the Roaring Twenties in the United States followed by the Wall Street Crash of 1929 and the Great Depression, the dot-com bubble involving new electronic technology and the Internet in the late 1990s, the American subprime lending boom in the 1990s and early 2000s, followed by the subprime mortgage crisis of 2006.

Famines have immediate health impacts sufficient to kill people, and babies in utero during famines suffer significant poor health for the rest of their lives. The classic example being those babies in utero during the Dutch famine of 1944/45. Known as the Dutch Hunger cohort, Lumey found that this cohort had a 10% higher mortality rate than those born before or conceived after the event. They suffered from higher rates of obesity, diabetes, and schizophrenia. Bielski brothers, partisans who survived and fought in Ukraine in WWII, were not considered successful before WWII. However, their skill sets proved suitable, whereas business men they helped survive had few or no forest survival skills, although doctors, nurses, and clock makers (they became gunsmiths) were highly valued. Those who survived the sinking of the Titanic were predominately kids, who survived in part because they suffered little or no psychological stress.

The Medieval Black Death ushered in new royal lineages and new labor power structures. The church's power was weakened and, in some cases, the social roles it had played were taken over by secular groups. Social mobility increased, as depopulation further eroded the peasants' already weakened obligations to remain on their traditional holdings. The sudden shortage of cheap labor provided an incentive for landlords to compete for peasants with wages and freedoms, an innovation that, some argue, represents the roots of capitalism, and the resulting social upheaval "caused" the Renaissance, and even the Reformation. Nicholas suggests a 150-year cycle from poor-to-rich-to-poor or clogs to clogs in three generations is evident in Britain since 1850 and suggests that the meek will inherit the earth if the world goes through a global collapse.

References

British Ecological Society. "Mowing urban lawns less intensely increases biodiversity, saves money and reduces pests." *ScienceDaily*. ScienceDaily, 19 December 2019. www.sciencedaily.com/releases/2019/12/1

Costanza, R., d'Arge, R., De Groot, R., Farber, S., Grasso, M., Hannon, B., Limburg, K., Naeem, S., O'neill, R. V., Paruelo, J., & Raskin, R. G. (1997). The value of the world's ecosystem services and natural capital. *Nature, 387*(6630), 253–260.

Davis, M. (2018). Eat a 'flexitarian'diet to help stop climate change. *Big Think*. https://bigthink.com/surprising-science/flexitarian-diet-climate-change. Accessed 11 Jan 2020.

Ekamper, P., Van Poppel, F., Stein, A. D., & Lumey, L. H. (2014). Independent and additive association of prenatal famine exposure and intermediary life conditions with adult mortality between age 18–63 years. *Social Science & Medicine, 119*, 232–239.

Emery, I., & Molidor, J. (2019). *Catering for the climate: How earth friendly-menus at events can save the earth.* https://www.takeextinctionoffyourplate.com/.pdfs/environmental_catering_report_catering_to_the_climate_final_report_2019.pdf

9 Geyer, R., Jambeck, J. R., & Law, K. L. (2017). Production, use, and fate of all plastics ever made. *Science Advances, 3*(7), e1700782. https://doi.org/10.1126/sciadv.1700782.

González, A. D., Frostell, B., & Carlsson-Kanyama, A. (2011). Protein efficiency per unit energy and per unit greenhouse gas emissions: Potential contribution of diet choices to climate change mitigation. *Food Policy, 36*(5), 562–570.

Hernandez-Aguilera, J. N., Conrad, J. M., Gómez, M. I., & Rodewald, A. D. (2019). The economics and ecology of shade-grown coffee: A model to incentivize shade and bird conservation. *Ecological Economics, 159*, 110–121.

Intergovernmental Panel on Climate Change. (2018). *Global warming of 1.5° C: An IPCC special report on the impacts of global warming of 1.5° C above pre-industrial levels and related global greenhouse gas emission pathways, in the context of strengthening the global response to the threat of climate change, Sustainable Development, and Efforts to Eradicate Poverty*. Intergovernmental Panel on Climate Change. URL: icpp.ch.

Janszen, E. (2010). *The postcatastrophe economy: Rebuilding America and avoiding the next bubble*. New York: Penguin.

Knapp, A. K., Blair, J. M., Briggs, J. M., Collins, S. L., Hartnett, D. C., Johnson, L. C., & Towne, E. G. (1999). The keystone role of bison in North American tallgrass prairie: Bison increase habitat heterogeneity and alter a broad array of plant, community, and ecosystem processes. *Bioscience, 49*(1), 39–50.

Kunstler, J.H. (2007). *The long emergency: Surviving the end of oil, climate change, and other converging catastrophes of the twenty-first century*. New York: Open Road+ Grove/Atlantic.

Lehrburger, C., Mullen, J. and Jones, C.V. (1991). *Diapers: Environmental impacts and lifecycle analysis* (No. 677.21 L524d). Massachusets, US: sn, 1991.

Leitner, R. (2014). British students force end of Barclays Bank's investments in South African apartheid 1969–1987. Global Nonviolent Aection Database, nvdatabase.swarthmore.edu. Accessed 12/25/2019.

Leopold, A. (1989). *A Sand County almanac, and sketches here and there*. Outdoor Essays & Reflections. Albany, NY.

Luntz, S. (2020). *A slight change in altitude could slash flying's climate cost.* 12 Feb 2020 iflscience. https://www.iflscience.com/environment/a-slight-change-in-altitude-could-slash-flyings-climate-cost/ Albany, NY

Meadows, D.H., Randers, J. and Behrens III, W.W. (1972). *The limits to growth: a report to the club of Rome*.

McNew, L. B., Winder, V. L., Pitman, J. C., & Sandercock, B. K. (2015). Alternative rangeland management strategies and the nesting ecology of Greater Prairie-Chickens. *Rangeland Ecology & Management, 68*(3), 298–304.

Monnet, J. (1978), *Memoirs*, London: Collins.

Moran, M. D. (2014). Bison grazing increases arthropod abundance and diversity in a tallgrass prairie. *Environmental Entomology, 43*(5), 1174–1184.

Moran, M.D. (2019). Once Doomed, America's recovering Bison are saving US ecosystems. *The Conservation*, January 19, 2019. https://www.inverse.com/article/52602-bison-are-back-and-that-benefits-many-other-species-on-the-great-plains

Moyer, E. J., Hulon, M. W., Butler, R. S., & Williams, V. P. (1992). *Lake Tohopelakiga muck removal project* (pp. 18–25). In: Proceedings of the first annual Southeastern Lakes management conference, 19–21 March 1992. Kennesaw State College, Marietta, Georgia.C.E. Watkins, H. McGinnis, and K.J. Hatcher, Editors.North American Lake Management Society, Alachua.

Nicholas, T. (1999). Clogs to clogs in three generations? Explaining entrepreneurial performance in Britain since 1850. *The Journal of Economic History, 59*(3), 688–713.

O'Sullivan, O. S., Holt, A. R., Warren, P. H., & Evans, K. L. (2017). Optimising UK urban road verge contributions to biodiversity and ecosystem services with cost-effective management. *Journal of Environmental Management, 191*, 162–171.

Reddy, R. L., Reddy, V. S., & Gupta, G. A. (2013). Study of bio-plastics as green & sustainable alternative to plastics. *International Journal of Emerging Technologies and Advanced Engineering., 3*(5), 82–89.

Rifkin, J. (2013). *The European dream: How Europe's vision of the future is quietly eclipsing the American dream*. John Wiley & Sons.

Rizvi, S., Pagnutti, C., Fraser, E., Bauch, C. T., & Anand, M. (2018). Correction: Global land use implications of dietary trends. *PLoS One, 13*(10), e0205312. https://doi.org/10.1371/journal.pone.0205312.

Saha, N., Mollah, M.Z.I., Alam, M.F. and Rahman, M.S. (2016). Seasonal investigation of heavy metals in marine fishes captured from the Bay of Bengal and the implications for human health risk assessment. *Food Control*, 70, 110–118.

Salcito, K. (2015). *Why cloth diapers might not be the greener choice, after all. Washington Post*, May 8, 2015. Accessed Oct 17, 2018.

Sarraf, M., Stuer-Lauridsen, F., Dyoulgerov, M., Bloch, R., Wingfield, S. and Watkinson, R. (2010). *The ship breaking and recycling industry in Bangladesh and Pakistan*. Washington, DC.

Shirer, W.L. (1990). *20th Century Journey: A native's return*, 1945–1988 (Vol. 3). Simon and Schuster.

Stiglitz, J. E. (2010a). *Freefall: America, free markets, and the sinking of the world economy*. New York; London: WW Norton & Company.

Stiglitz, J. E. (2010b). Moral Bankruptcy. Why are we letting Wall Street off so easy? Politics, Mother Jones, January/February 2010.

Teoh, R., Schumann, U., Majumdar, A., & Stettler, M. E. (2020). Mitigating the climate forcing of aircraft contrails by small-scale diversions and technology adoption. *Environmental Science & Technology, 54*, 2941.

Thunberg, G. (2019). U.N.'s climate action summit, NYC. Sept 23.

Trigger, B. G. (1998). *Sociocultural evolution: Calculation and contingency*. Oxford: Wiley-Blackwell.

Turner, G.M. (2008). A comparison of The Limits to Growth with 30 years of reality. *Global environmental change*, 18(3), 397–411.

Wang, Y., & Akbari, H. (2016). The effects of street tree planting on urban Heat Island mitigation in Montreal. *Sustainable Cities and Society, 27*, 122–128.

Watson, C. J., Carignan-Guillemette, L., Turcotte, C., Maire, V., & Proulx, R. (2019). Ecological and economic benefits of low-intensity urban lawn management. *Journal of Applied Ecology, 57*, 436.

Wegener, W. E., & Williams, V. P. (1974). *Fish population responses to improved lake habitat utilizing an extreme drawdown*. Proceedings 27th annual conference of the Southeastern Association of Fish and Wildlife Agencies (pp. 663–674). Hot Springs, Arkansas.

Supplementary Information

M. R. Welford, R. A. Yarbrough, *Human-Environment Interactions*,
https://doi.org/10.1007/978-3-030-56032-4

Index

A

B

C

D

E

F

G

H

I

K

L

M

N

O

P

Made in the USA
Monee, IL
23 September 2025

30039892R00144